TRAVEL AND WONDER IN THE EARLY MODERN WORLD

This collection of chapters explores wonder in the context of early modern travel and travel writing, offering multifaceted and novel interpretations of the problematic relationship between a traveller and their unfamiliar environment through various geographical, chronological and thematic lenses.

Exploring representations, descriptions and uses of the unfamiliar, the contributors discuss and elucidate rhetorical, epistemological, religious, colonial, materialistic and emotional aspects of wonder in the early modern world. They study European travellers and their texts, as well as descriptions of wonder within the Muslim world and reactions to the unfamiliar reported by Muslim travellers in Europe. The collection ranges from travellers' descriptions of wonder to an analysis of wonders that have travelled. With its focus on "wonder" in the context of early modern travel, this volume fills a significant gap in research, shedding new light on the history of intercultural encounters and on the processes of learning about the world and our place in it.

The book is aimed at both academic and non-academic readers, for experts who study early modern history and travel writing, and for lay readers who are curious about the history of travel and about past conceptions of the world and foreign cultures.

Jaska Kainulainen is a Docent of the History of Ideas at the University of Helsinki, Finland. He is the author of *Paolo Sarpi: A Servant of God and State* (2014) and *Early Jesuits and the Rhetorical Tradition* (2024), along with multiple articles and book chapters.

TRAVEL AND WONDER IN THE EARLY MODERN WORLD

Representations, Descriptions and Uses of the Unfamiliar

Edited by Jaska Kainulainen

Routledge
Taylor & Francis Group
LONDON AND NEW YORK

Designed cover image: The lighthouse at Alexandria, Egypt. Handcolored copperplate engraving from Robert von Spalart's "Historical Picture of the Costumes of the Principal People of Antiquity and of the Middle Ages," Metz, 1810. Florilegius / Alamy Stock Photo.

First published 2026
by Routledge
4 Park Square, Milton Park, Abingdon, Oxon, OX14 4RN

and by Routledge
605 Third Avenue, New York, NY 10158

Routledge is an imprint of the Taylor & Francis Group, an informa business

British Library Cataloguing-in-Publication Data
A catalogue record for this book is available from the British Library

ISBN: 978-1-032-82909-8 (hbk)
ISBN: 978-1-032-82910-4 (pbk)
ISBN: 978-1-003-50692-8 (ebk)

DOI: 10.4324/9781003506928

Typeset in Times New Roman
by SPi Technologies India Pvt Ltd (Straive)

CONTENTS

CONTRIBUTORS

Ipek Bozkaya is a PhD candidate in history and a research associate at the Johannes Gutenberg University in Mainz, Germany.

Antoine Eche is an associate professor at the department of English, languages and cultures at Mount Royal University in Calgary, Canada.

Hana Ferencová is an assistant professor of early modern history at Palacký University Olomouc, Czech Republic.

María Juliana Gandini is an assistant researcher at the National Scientific and Technical Research Council (CONICET) and an adjunct professor of history at the National University of Luján, Argentina.

María Isabel Gaviria is a PhD candidate in the Romance studies at Heidelberg University, Germany.

William Glover is a PhD candidate in English literature at Boston University, United States.

Jaska Kainulainen holds the title of docent in the history of ideas at the University of Helsinki, Finland.

Mahmoud Abdelhamid M. A. Khalifa is an assistant professor of cultural studies and literary criticism at the Department of English, Faculty of Al-Alsun (languages), in Hurghada University, Hurghada, Egypt.

Susanna Lahtinen is a PhD candidate in cultural history at the University of Turku, Finland.

Dinu Luca is a professor of European studies at the National Taiwan Normal University in Taipei City, Taiwan.

Ioana Manea is an associate professor of French literature at Ovidius University of Constanţa, Romania.

ACKNOWLEDGEMENTS

The chapters of this book were originally drafted as papers for a conference held in April 2023 at the Centre for Renaissance and Early Modern Studies (CREMS) in the University of York. The conference "Travel and Wonder, 1450–1750" was generously funded by KONE Foundation. Besides the funding body, I would like to thank Professor Simon Ditchfield, who, as the head of CREMS, kindly offered to host the event and helped to complete numerous organisational tasks. For the same reasons, I thank his followers as the head of CREMS, Professor Kevin Killeen. Importantly, much could have gone awry without the administrative help from Megan Russell. I would also like to thank our keynote speakers Joan-Pau Rubiés and Lauren Working for their wonderful contributions to the event. In a similar vein, thank you to Riyaz Chenganakkattil, Dominic Gavin, Orazio Gnerre, Lorenz A. Hindrichsen, Emilie Murphy, Emily Stevenson and Soile Ylivuori.

Jaska Kainulainen

1

INTRODUCTION

Ways of Seeing and Reporting Wonder

Jaska Kainulainen

"Seeing comes before words", the art critic and novelist John Berger writes at the beginning of his essay *Ways of Seeing*. While a child looks and recognises without words, an adult attaches words to images, although, Berger continues, "the relation between what we see and what we know is never settled".[1] This is particularly true in the context of travel, when preconceived ideas and knowledge of familiar things clash with the unknown, with the (seemingly) strange things that we perceive in a new and unfamiliar environment. This collection of essays explores early modern travellers' experiences of what they labelled as wonders, focusing on the ways in which they perceived, conceptualised and reported such phenomena in their travelogues. The underlying idea of the book is that if wonder indicates rarity and mysterious qualities, travel documents are a particularly suitable source material for a study about the nature and the uses of the wondrous. Past travel writers were, in a sense, creators of wonders as their published descriptions of the curiosities of distant regions were disseminated among the reading public. These descriptions echo the unsettling process of finding words for images, the task of naming things which had no name, the desire to explain to others the full impact of something that seemed like a dream. In the era of discoveries and rapid expansion of the known world, "wonder" became a short-hand for the impenetrable, incomprehensible, mysterious, and, like a child, the travel writer called a variety of things by the same name.

Images are powerful and relational—who sees and who shows, what do we decide to look at and what does the Other show us? Do we see something accidentally or purposefully, are we innocent when we look, or is our gaze directed towards prejudiced ideas, eager to prove them right? Perhaps everything that a

DOI: 10.4324/9781003506928-1

traveller sees and writes about is imperfect, flawed and distorted. Perhaps past travellers' fumbling amidst deceptive vistas and ambiguous meanings only produced histories of misunderstanding, misinterpretation, bigotry and confusion. If this is true, how should we read such communications from the past? Are wonders born from a persistent failure to understand?

Like images, wonders too are powerful and relational. There is no wonder without a background, without the observer's mentally constructed cultural framework against which the wonder can shine. When reading travel writers' descriptions of wonder it is therefore important to acknowledge that every observer looks from a certain position and that there is a perspective behind every travelogue and wonder. Is it the case that the fuller the view, the smaller the miracle? Do wonders emerge (only) when the perspective does not allow a sufficient viewpoint?

What does seeing tell us about our place in the world? Is everything that we see marked by fear or desire? Or disgust, awe, admiration and wonder? If this is the case, are travel documents primarily attempts to process emotional responses to the unfamiliar? Do such emotional responses issue from a tension between the unfamiliar surroundings and the traveller's predetermined, habitual idea of his or her identity and place in the world? Finally, is not any travel document just a traveller's exploration of the self? Even if self-exploration was not the explicit aim of the writing process, it certainly is the case that travel writing can be read as documentation which reveals aspects not only of the writer's identity and cultural background, but also of their fears, desires and prejudices. This interpretative work no doubt relies on the reader's familiarity with the historical context and their sensitivity to the changing registers in which the writer describes the journey and the unfamiliar. In reading travel documents, it is therefore important to bear in mind aspects of relativity, to establish the traveller's position with regard to the object that is being described in the text, and to understand the movements of light—what is in spotlight, what remains in shadows, how full or dim is the view.

The problem of ungrammaticality

As a young man, the novelist V.S. Naipaul arrived from Trinidad to rural Wilthsire in the United Kingdom and later wrote of that moment that the English countryside was so foreign to him that he failed to understand what he saw. "I saw what I saw very clearly", he writes, "but I didn't know what I was looking at". In his words, Naipaul felt "unanchored and strange", and he saw "things slowly; they emerged slowly".[2] If this sort of myopia applies to all travellers and if all travel writing is born from confusion and faulty understanding, what is it that we are dealing with when we read travel literature? What are the questions that we should be asking of such texts? In the case of early modern travel writing, our reading is further complicated by the cultural and linguistic alienation caused by the temporal distance between us and the past travellers who composed their observations about the unfamiliar.

In some sense, travellers become "ungrammatical" to borrow the term that John Berger uses to describe primitive artists who have not learnt the "pictorial grammar" of professional art (another word Berger uses is "clumsy", because primitive artists have not learnt the technical skills of the convention).[3] Even when the language was a familiar one, say, Latin, early modern travellers may have struggled to understand a conversation due to the strangeness of the context that otherwise perfectly familiar words were part of.[4] This sort of ungrammaticality is also referred to in Naipaul's autobiographical narrative when he describes a conversation between him and a young man who was using a tractor to break a field of grass, concluding that "even when what the young man said was made clear to me, it didn't make sense".[5] In a similar vein, when Naipaul looked at English gardens, he only saw colour, "barely able visually to disentangle one plant from another". This was because "in England I was not a gardener", he concludes. While familiar with the flowers and plants of Trinidad, when surrounded by English flora, Naipaul's vision became impaired. With his previous familiarity with plants in Trinidad, however, he was able to resort to "instinctive knowledge" and, slowly, to learn to see the plants of Wiltshire. This, he remarks, "was a bit like learning a new language".[6]

The endeavour to overcome ungrammaticality by attaching words to images is a first step towards dismantling and demystifying what is strange and wondrous. In a way, by describing the wonders of the world, past travel writing both created the wonders and commenced the gradual process of obliterating them. The concomitant production of pictorial representations of unfamiliar places and cultures was a compelling means to illustrate the wonders of the world to those who did not travel. Wonders are resilient, however: it is difficult to explain them, either because they are incomprehensible or because they elude definitions or because it is prohibited to explain them. Guardians of a religious doctrine, for example, might defend wonder's mysterious qualities from scientific revelations or what the guardians might consider a wrong kind of interpretation. Furthermore, not everyone is able to visit a distant wonder, to see it with their own eyes and to study it until the mystique starts to dissolve. Wonders tend to survive and spread in stories and pictorial representations.

Marlow, the narrator in Joseph Conrad's *The heart of darkness* (1995), describes his ship's slow journey on an African river in terms which associate wonder with insufficient knowledge and the failure to read and understand unfamiliar surroundings: "watching a coast as it slips by the ship is like thinking about an enigma". The foreign landscape was there, "smiling, frowning, inviting, grand, mean, insipid, or savage, and always mute with an air of whispering, Come and find out".[7] At the beginning of his story, Marlow reminisces how, as a boy, he used to look at the map of Africa that fascinated him "as a snake would a bird—a silly little bird".[8] Conrad gives a sinister hue to Marlow's fascination with the distant, alien region. The curiosity of the white man is linked to colonialism and greed, while the young boy's bygone, innocent fascination with the strange and exotic turns into a "weary pilgrimage amongst hints for nightmares". The "dance of death and trade" that Marlow witnesses amidst "formless coast" and "contorted mangroves"—in an atmosphere of "overheated catacomb"—leaves him ungrammatical. He states that "nowhere did we stop long enough to get a particularised impression, but the general sense of vague and oppressive wonder grew upon me".[9] The further Marlow travels, the more obscure his narrative becomes.[10] At this point, it is important to bear in mind that Marlow's story was created by a seasoned traveller: prior to his life as a writer, Conrad was, for a period of approximately fifteen years, a sailor who travelled extensively on the Pacific and whose visit to the Congo in 1890 inspired his subsequent writing.

There is a connection between Marlow's gloomy ruminations about the "oppressive wonder" that was unravelling in front of him and Conrad's relationship with writing. Using English rather than his native Polish, Conrad once remarked that he had to "work like a coal miner in his pit quarrying all my English sentences out of a black night".[11] In both Marlow's journey and in Conrad's literary effort to create that journey, there is a sense of despair that derives from the ephemeral randomness of "words that disappear in a spectral opacity before they may be briefly embodied in any form at all", to quote Andrew Rubin.[12] While both the traveller and the writer sought to make sense of what was in front of them, words (with their fragile, invented connection to reality) were the means, no matter how inadequate, to achieve that. As Edward Said puts it, Marlow "acknowledges the tragic predicament of all speech" when he pauses to lament the difficulty of telling his story, which to him felt like trying to "tell you a dream". This was a "vain attempt, because no relation of a dream can convey the dream-sensation". Marlow adds that to convey others "the life-sensation of any given epoch of one's existence", whatever "makes its truth, its meaning", is "impossible", because "we live, as we dream—alone".[13]

Perhaps we can see in Marlow/Conrad the merging of the rationalistic and empirical traditions, the Cartesian writer who invents through cognitive processes and the Baconian sailor who learns by doing. Denis Diderot (1713–84) made this distinction pungently when he mused that "the contemplative man" is a "sedentary being; the traveller is either ignorant or a liar". While the former was, in Diderot's view, "contemptuous of the minute details of experience", the latter—"he who experiences"—was "almost always without genius".[14] This clearly demonstrates the Enlightenment philosopher's distrust in travel writers' ability to grasp and report the unfamiliar sights before them. Perhaps Conrad regarded fiction as the best way to process his experiences as a traveller.

At times, travel writers use the sort of ungrammaticality that blurred their judgement in a new and unfamiliar environment as a rhetorical means to amplify the wonderful strangeness of the situation they found themselves in. Upon her arrival in Antigua in December 1774, Janet Schaw (1731–1801)—the author of the travel diary *Journal of a lady of quality*—observed that "everything was as new to me, as if I had been but a day old". She praises the richness of the colours and perfumes of the West-Indian flora (perhaps commonplace to Naipaul), which, she adds self-consciously, "mock the poor imitations to be produced by art".[15] The Indian diplomat Mirza Sheikh I'tesamuddin (ca. 1730–1800) in turn describes the effect of a dance party he was invited to while visiting London in 1765 as something that rendered him unable to "distinguish between the brightness of the lamps and the splendour" of the ladies present at the party. He was left "speechless with admiration" and "overwhelmed by the glory of Allah's creative power".[16]

Whatever the nature of the wonder—natural, artificial, spiritual—any description of it is man-made and, accordingly, the chapters of this book explore man-made, written representations of wonder. Ultimately, when we read descriptions of wonder in a travelogue, we read words that amount to the writer's endeavour to make or construct that wonder through mimesis. Edward Said emphasises the significance of Giambattista Vico's (1668–1744) *verum factum* principle—the idea that we only know what we have made—not only to his own interpretation of orientalism but also to Erich Auerbach's understanding of the relation between writing and reality, a relation marked by mimesis and representation.[17] In Auerbach's classic work *Mimesis: the representation of reality in Western literature* (1946), literary text represents reality through mimesis (imitation or interpretation). In other words, the human mind has made the reality that we read about. In his *La scienza nuova* (1725), Vico distinguishes between three different ages of mankind, the divine, the heroic and the human, which all had their own literary or linguistic system. The language of the divine age was developed by people who were called "poets" because, as Vico points out, the Greek for "poet" also meant "creator". The poets were guided by the "robust ignorance" of the age, the "ignorance of causes" which in turn was the "mother of all wonders", and their language was marked by the sublime and "evocative of wonder".[18] According to Vico, curiosity too was born from ignorance—wonder and curiosity were siblings, while science was the daughter of the latter—and whenever people witnessed a "strange natural phenomenon" such as a "comet", curiosity drove them to wonder its meaning.[19]

Vico postulated a past when there was a "mental language common to all nations", a language which not only "understands uniformly the substance of things that are feasible in human social life" but also explains that substance with as many "diverse modifications" as the things have "diverse aspects".[20] It has been pointed out that even as the world depicted in Vico's elucidation of the past became increasingly complex, it was still his position that "man's thought and mental language of images developed common broad metaphors which could provide the basis for understanding among people".[21] While the English-language scholarship on the history of travel and travel writing is predominantly focused on the experiences of European travellers and on their descriptions of the strange, exotic and marvellous, it goes without saying that, in line with Vico's idea of common metaphors, non-European travellers experienced similar confusion before strange phenomena and shortage of words to describe them. In such situations, the notion of wonder—whichever language or word denoted it—functioned as Vico's common metaphor, as a universal, human attempt to address the unfamiliar.

Perspective and perception

In early modern travel writing, new and unfamiliar phenomena appear wondrous either because they are somehow veiled or disproportionate when compared to similar but familiar things. On the other hand, being wondrous can also mean that a phenomenon is simply beyond comparison. Discussing themes such as proportion, comparison and familiarity, the Italian polymath Leon Battista Alberti (1404–72) writes in his treatise *On painting* that while the surface of an object remains unaltered, its qualities "appear changed to the observers either with the changing of position or with the changing light". He also notes that reflection can make "the faces of those who walk on meadows appear greenish".[22] While what we see depends on our position and on the quality of light, the process of understanding what we see depends on comparison. "Ivory and silver are white", Alberti writes, "but if compared to a swan or to snow-white linen cloths, they appear somewhat pallid". According to him, comparison must be made, "first of all, among the best-known things". Quoting Protagoras's famous maxim, he adds that since, of all objects, the human figure is "the best known to man", it follows that "man is the model and the measure of all things".[23] Things that appear disproportionate to the human figure and to other "best-known things" seem strange and monstrous. To avoid this, Alberti writes that "in dressing one first needs to know the nude, which we then envelop encircling it with clothes" and when painting a nude "one needs first to arrange the bones and the muscles" which will then be covered "with flesh and skin".[24] His book is a classic introduction to what Berger called "pictorial grammar" in the West, and its aim is to guide aspiring painters to avoid disproportionality and to complete admirable works of art through methodical understanding of both their discipline and the objects they want to paint. For Alberti, the wondrous was located in the classical understanding of harmony and proportionality, it was rational, geometrical and mathematical. What Alberti's book sought to achieve was the overcoming of ungrammaticality, the plight that conditioned early modern travellers' expositions of the strange and the unfamiliar.

Perhaps descriptions of wonder in travel literature are comparable to what Alberti would have considered a clumsy painting—a product of a flawed understanding of both the object and the grammar with which to describe it correctly. On the other hand, it might be that eyesight is not as reliable and important as, for example, William Ivins Jr. implies in his 1938 essay "On the rationalization of sight". Perhaps eyesight is treacherous and deceitful in the manner described by the English clergyman George Hakewill in his *The Vanitie of the eie* (1608). In early modern theological discourse devil tricked the eyesight by performing wondrous acts to evil ends—as acknowledged, for example, by Hakewill—although, in the end, the devil's powers were limited to the natural world and its wonders (*mira*), leaving true miracles (*miracula*) exclusively to god.[25] It is undeniable that for many early modern travellers there was a metaphysical, supernatural, spiritual dimension to the world. What can eyesight disclose of such a dimension? What use is there for a theory of perspective in such a world?

The French philosopher Maurice Merleau-Ponty asserted that "a child perceives before he thinks".[26] His phenomenological interpretation of seeing challenged the idea that perception would be an "intellectual act", which would "grasp the object either as possible or necessary". By "possible", he means that when we do not see all parts of an object, we should represent to ourselves the parts which we cannot see (the other side of a lamp, for example). However, to represent to ourselves the side of the lamp that is outside our vision means that we conceive of that invisible side of the lamp as a representation and not as something actually existing. "What is represented", he writes, "is not here before us", and "I do not actually perceive it" which means that "it is only a possible". By "necessary", Merleau-Ponty means that if we look at a cube, "knowing the structure of the cube as it is defined in geometry", we can anticipate what the other side of it will look like. Therefore, he asserts, "I would know the unseen side as the necessary consequence of a certain law of the development of my perception". However, while "it is true that the lamp has a back" and the "cube has another side", the formula "it is true" does not correspond to what is actually perceived. This is because "perception does not give me truths like geometry but presences".[27]

According to Merleau-Ponty, the objects that we perceive are neither representations nor results of our intellectual synthesis, but, instead, they are real and they result from "an indefinite series of perspectival views" each of which gives us the object, although none of them does it "exhaustively". This "deformed" or imperfect perception is the "price of its being real". Merleau-Ponty adds that "the places in which I find myself are never completely given to me" because the perceived objects always "recede beyond their immediately given aspects".[28] However, the imperfection of our perception should not make us abandon reflection. What Merleau-Ponty is saying is that, when reflecting, we should be aware of the development or birth of our thinking. Its origins lie in perception. This realisation has consequences on how we develop (and perceive) our own thinking. What and how do we perceive? What is our perspective? What is our positionality?

How did early modern travellers deal with the invisible, with the other side of the lamp or the cube? How did they explain the elusive aspects of the objects they perceived? Focusing on past travellers' perceptions and descriptions of wonder, the chapters of this book broach—in different ways—issues related to Merleau-Ponty's notions of "possible" and "necessary", to perception, representation, credibility and, ultimately, reality. It seems safe to say that theories of perspective and perception allow and acknowledge an invisible, unknown side to every object or phenomenon. It is that unseen side which is then, more or less freely, pieced together by representations of the possible or by inferences based on the necessary. Following these observations on perspective and perception, it seems that wonders grow either from the unseen or, when seen, from a perceived disproportionality.

The wonder and the sublime

The connection between wonder and the sublime is made evident in Nicholas Boileau's 1674 translation of Longinus's *Peri hypsous* (On the sublime). In his translation, Boileau writes that the sublime is "strikingly extraordinary and marvellous in discourse", something that makes a piece of writing "elevate, ravish and transport". Besides the extraordinary and the marvellous, Boileau lists "the surprising" as a further quality of the sublime.[29] According to Robert Doran, Boileau's preface to his translation introduced Longinus's work to the "European intellectual culture" and "made *le sublime* a fashionable term".[30] In his treatise, Longinus (or Pseudo-Longinus, an unidentified first-century author) writes that "invariably what inspires wonder, with its power of amazing us, always prevails over what is merely convincing and pleasing". What is more, in contrast to rational speech that persuades with its gradually emerging facts, the sublime—like wonder—takes the audience by surprise: "a well-timed flash of sublimity shatters everything like a bolt of lightning and reveals the full power of the speaker at a single stroke". According to Longinus, the effect of the sublime in a speech is not to persuade the audience but "rather to transport them out of themselves".[31] Wonders—natural or man-made—have similar impact on travellers: they arouse amazement and awe. They do not persuade but, instead, they demand the traveller to capitulate and desist all attempts to understand and explain their nature, meaning and causes. Rather than appealing to cognitive processes, wonder circumvents reason and targets the senses with a profusion of strangeness.

As the full title of Nicholas Boileau's translation of Longinus suggests, early modern writers considered "wonder" and "sublime" synonymous. This is also evident in Philip Francis's 1746 translation of Horace's *Ars poetica*: the former rendered Horace's sentence "I compose wonderful poems" into "I can write the true sublime".[32] Edmund Burke, in turn, associated the sublime with fear and terror, arguing that the latter was "the ruling principle of the sublime". He then replaces "sublime" with terms more obviously relating to "wonder" and argues that this association (between fear and the sublime) is present in many languages as "they frequently use the same word" to signify "the modes of astonishment or admiration and those of terror". In Greek, Burke asserts, "*thámbos*" stands for "either fear or wonder", the Latin "*attonitus*" (thunderstruck, stunned) denotes fear and astonishment, and "the French *etonnement* and the English astonishment and amazement" indicate "the kindred emotions which attend fear and wonder".[33]

It was the unknown and unfamiliar nature of foreign lands, customs and phenomena that provoked fear and wonder in a traveller. In *The heart of darkness*, Conrad's narrator digresses into an imaginative speculation about ancient Romans arriving at the English shore with "nothing but Thames water to drink", forced to "live in the midst of the incomprehensible", which, however, "is also detestable" and "has a fascination, too". The unintelligible, alien landscape had the power to provoke a wide range of emotions, both positive and negative ones.

Alberti writes that an "agreeable and rich" painting can "attract for a long time the eyes" of a spectator "with a certain sense of pleasure and emotion". What brings pleasure in a painting is "richness itself and a variety of objects". In particular, he adds, "new and extraordinary things always delight" because "they differ from those old and customary things, so the mind feels pleasure exceedingly in every variety and abundance of objects".[34] In other words, new and unfamiliar objects and phenomena bring pleasure because they are not trite, common and obvious. In this sense, wonder is the opposite of banal. Also this aspect of wonder—its ability to give pleasure through novelty—implies a challenge to cognitive processes. Janet Schaw was "resolved to enter into no particulars" about West-Indian plants and flowers because "the beauty, the novelty, the ten thousand charms that this scene presents to me, confuse my ideas".[35]

Perhaps it takes a certain mindset to register marvels. Perhaps some travellers were happy not to explore the causes behind wondrous phenomena, not to unravel the mystery behind what was strange and unfamiliar, but surrendered themselves to the raptures of wonder. Then there were travellers who seemed to have remained impervious to everything unfamiliar, plodding ahead with a focus strictly on whatever their mission was. In 1620, Giles Hobbs travelled from Moscow to Spahan in Iran, a journey no doubt filled with both hardships and wonders. His writing is, however, remarkably impassive and void of both curiosity and observations about the surroundings. When mentioning his "passage on the Volga", Hobbs dryly adds that he "held nothing worthy notice except the taking the sturgeon, and the making the caviar and isinglasse".[36] Although Hobbs was constantly surrounded by new and strange landscapes, plants, customs and languages, nothing seemed to astonish him. In his insipid register, the reference to sturgeon is a passionate remark, one that reveals his desire for caviar. Albeit in a muted tone, it was the sudden familiarity amidst the unknown that sparked a more animated response. One wonders if Hobbs' silence about all the wondrous and strange things that he must have encountered on his route from Moscow to Spahan simply testifies to his exhaustion caused by the over-abundance of foreignness, sensory stimuli and the terrors of the sublime.

In such instances familiarity offers consolation, although not necessarily without a spell of melancholy. In *The heart of darkness*, Marlow recalls how he discovered a book inside a "hut of reeds" and how that book—"something unmistakably real"—made him forget the surrounding jungle, the unreal, menacing, dream-like "night of first ages". When he had to put the book aside and return to the steamboat, it was like tearing himself away "from the shelter of an old and solid friendship".[37] In a letter composed during her journey to Scandinavia, Mary Wollstonecraft writes how her "eye was attracted by the sight of some heart's ease that peeped through the rocks". She regarded the familiar flower as a good omen and, while noting its presence in Shakespeare, concludes that the sight of the flowers evoked complex emotions and associations that "suffused" her eyes. At the same time, however, her young daughter experienced "unmixed gaiety" because she "found a few wild strawberries".[38]

Wonder, science and the credibility of travel writing

The Oxford English Dictionary defines the noun "wonder" in the first place as "a marvellous object; marvel, prodigy" and, secondly, as "marvellous character or quality". The dictionary also describes "wonder" as a "deed" or "event" which has been "brought about by miraculous or supernatural power; a miracle", while the obsolete use of the word in the sense of omen or portent referred to "an extraordinary natural occurrence".[39] What these descriptions have in common is the fact that wonder has the power to evoke astonishment, surprise and admiration in those who witness it. It also seems that the astonishment and admiration that are brought about by wonder are ultimately caused by the inexplicable, mysterious nature of the phenomenon. Wonders are, by their nature, impervious to explanation and analysis. As Mary Baine Campbell has pointed out, "the relation of wonder to knowledge is crucial but largely oppositional", wonder being "a form of perception" mainly associated with "innocence", with children and the uneducated.[40] At the same time, however, wonder has often functioned as the starting point of a scientific investigation. As Lorraine Daston and Katherine Park state in their classic study *Wonders and the order of nature, 1150-1750*, for early modern philosophers such as Robert Boyle, René Descartes and Francis Bacon wonder was "a goad to inquiry" and, in general, "ignorance of causes was traditionally a spur to wonder".[41]

In a similar vein, as we have seen, Giambattista Vico considered wonder and curiosity siblings who were birthed by ignorance, although he then identified the activity that was caused by curiosity as poetry, as an act of creating. In Vico's view, the "most sublime work of poetry is to give sense and passion to inanimate things" in the same way as children who pick up inanimate objects play with them as if they were alive.[42] Neither poets nor children are (easily) bound by rules or conventional views about grammar. They play and create in the same zone of ungrammaticality as travellers to unfamiliar regions do.

How can a traveller convincingly—or even comprehensibly—describe things which seem strange and unfathomable? How to do this without reducing everything to monosyllabic expressions of bewilderment, such as Marlow's exclamation "absurd!" when sensing the abyss between his narrative and his listeners' life (or the last words "the horror! the horror!" of the novel's enigmatic ivory trader, Kurtz)?[43] To convey anything more than such oracular utterances, the travel writer must first overcome the ungrammaticality which is forced upon him or her by the unfamiliarity of the environment, the nature, the culture, the language.[44] This is not an easy task, as implied by the fact that the ungrammaticality is often circumvented by formulations which state that something is too strange or marvellous for the writer to describe in words. Referring to "performances of music and dance" in Europe, I'tesamuddin concedes that "my pen lacks the ability not only to describe properly the many interesting things I heard and saw in such performances, but even to write a short panegyric".[45] Conversely, a couple of pages earlier I'tesamuddin gives the reader a detailed list of all things Indian he had seen at the British Museum, demonstrating his familiarity (grammaticality) with the objects by naming many of them in a language spoken in India.

It has been pointed out that early modern travellers to the New World "could find neither the proper words nor suitable Old World analogues for American facts", and this caused "strains on the one cultural tool" by which the traveller hoped to explain the marvels of the new continent to others.[46] When trying to portray Montezuma's palace to Charles V, Hernán Cortés had to admit that the building was "so marvellous" that it was "impossible to describe its excellence and grandeur", which is why he "shall not attempt to describe it at all". A market in Mexico City proved equally challenging for Cortés who first lists some familiar items but then gives up and states that due to the amount and variety of things at the market and "because I cannot remember many of them nor do I know what they are called I shall not mention them".[47]

Not only did the strangeness of the perceived phenomena challenge the traveller's ability to describe them, but depictions of such oddities tended to question the writer's credibility. The latter aspect of travel writing led the seventeenth-century Italian traveller Francesco Negri to inform his readers that he had "left out many things" either because he had not been able to confirm their veracity or because "they were true, but so incredible" that to write about them would have brought him shame.[48] Readers' scepticism about travel writers' descriptions of extraordinary phenomena was only augmented by the fact that it was usually difficult or impossible to find alternative sources to verify the account. This was particularly true with early travelogues such as Marco Polo's *Description of the world* (ca. 1298) and Ibn Battuta's *The rihla* (1355—or 756 in the Islamic calendar).[49]

In the age of discoveries and scientific revolution, philosophers started to emphasise the importance of empirical, sensory evidence and experimental procedures. They questioned excessive reliance on received wisdom (especially Aristotle) and, instead of old texts, advocated observation of nature as the key to improved understanding of the world. In his *Natural history of the Indians* (1590), José de Acosta criticised those who have written about the Americas without dealing with "the causes and reasons for those new things and natural wonders" and adds that these writers have failed to properly discuss the topic because they have been led astray by "philosophy formerly received and argued". What is more, their understanding of the local culture remained superficial, because "they did not know their language or because they did not bother to learn about their ancient history".[50]

A staunch critique of Aristotle, Francis Bacon writes in his *Novum organum* (1620) that "men have been held back from further progress in the sciences by the siren song of reverence for antiquity". In his view, this was particularly frustrating in an era when "by prolonged voyages and journeys (which have become prevalent in our times) many things in nature have been disclosed", and this in turn "could shed new light on philosophy". Bacon states that "it would be a disgrace to mankind" if "the limits of the intellectual globe were not pushed beyond the narrow confines of the ancients' discoveries".[51] The discovery of new geographic regions drastically altered the foundations of knowledge, generating doubt and ambivalence among early modern thinkers. Bacon writes that "to the human intellect reflecting on it, the fabric of the universe looks in its construction like a labyrinth, where we find everywhere so many blind alleys, such deceptions and misleading signs and such oblique and intricate convolutions and knots of nature". For Bacon the right path through the labyrinth consisted of inductive reasoning, the systematic observation of particular phenomena which then formed the basis for broader conclusions. "Our tracks must be guided by a clue", he writes, "and a sound policy must secure every step of the way right from the very perceptions of the senses".[52]

According to Wayne Franklin, Bacon regarded Columbus as a "symbol of bold modernity" and as his "hero as much as his metaphor". This interpretation of Columbus "provided travel with a philosophy which it had not previously had" while the traveller—who was free from the cultural context which in Bacon's view "warped the individual's sense of the natural world"—seemed to "enjoy a certain primary contact with the universe".[53] Bacon's admiration for

Columbus was based on the latter's achievements as an explorer whose "epic voyage across the Atlantic" was first doubted but "afterwards vindicated by his experiment".[54] For Bacon, Columbus's voyage was a scientific experiment that verified a hypothesis. As this introduction suggests, however, it is questionable whether a traveller, no matter how scientifically trained, can ever be totally free from the cultural context that "warps" his or her perception of the world.

Learning is hostile to wonder, even if the beginnings of such learning lay in curiosity. Bacon questioned the "amazement" caused by libraries and the "huge variety of books" in them because, in the end, they were full of "repetitions" and the reader would only be left with "wonderment at their poverty".[55] What Bacon advocated instead was learning based on observation and experiments. This had to be done with awareness of the "dullness, inadequacy and unreliability of the senses" and, he asserts, "all truer interpretation of nature is accomplished" with experiments, "where the sense judges only the experiment while the experiment judges nature and the thing itself".[56] It was Columbus rather than Aristotle who appeared to lead the way in the advancement of sciences and knowledge. According to Wayne Franklin, "words were the villain" of Bacon's new method because they misrepresented things and created "fantasy and speculation", which had their "only real existence in human speech".[57] Rather than merely speculating about geographical questions, Columbus set sail and expanded the boundaries of human knowledge with his bold experiment. While his discovery resulted in countless tales and descriptions of wonders, the underlying message of Bacon's new scientific method was that wonders could be explained with appropriate experiments. His belief in the scientific merits of travel was in stark contrast to authors such as Johann Boemus, whose *Omnium gentium mores, leges et ritus* (1520) based its ethnographic descriptions on established Latin sources rather than on vernacular texts written by contemporary travellers, whom Boemus referred to as "fraudulent vagrants" and "wandering liars".[58]

The hundred years that separate Boemus' *Omnium gentium mores* and Bacon's *Novum organum* witnessed not only the so-called scientific revolution, but, more importantly to this study, the publication of Giovanni Battista Ramusio's *Navigationi e viaggi* (1550–59) and Richard Hakluyt's *The principall navigations* (1589–1600), collections of travel documents which enabled, to a certain extent, the editors of the collections and their readers to compare and contrast information provided by individual texts. This kind of verification—which was not possible in the previous centuries—changed the status of travel writing and, in the words of Joan-Pau Rubiés, made the genre "authoritative in European intellectual culture".[59] However, as Diderot's disparaging remarks about travellers' reliability demonstrate, the novel verification methods and early modern travel writers' aspiration towards scientific style and veracity could not entirely free travel writing from charges of hyperbole and fabrication.

Referring to historians' contradictory portraits of Alexander the Great, Mirza Seikh I'tesamuddin writes that "the truth about him can be found only in the writings of other Greeks who knew him well". One should not rely on hearsay, he adds, because "there is a world of difference between what one hears and what one sees with one's own eyes".[60] In line with this observation, I'tesamuddin was sceptical of the claim made by some of his fellow countrymen (who were unfamiliar with both the seas and Alexander the Great) that Alexander had measured the range and depth of all the seas in the world. During his six-month-long journey from India to France, I'tesamuddin learnt the "vastness and terrifying power of the sea", the "unfathomable depth" of which was known to Allah alone. I'tesamuddin concluded that "the ocean is full of wonders", which were too numerous to be discussed in detail, but he briefly lists the flying fish, the whale, the sea cow, the hog fish and the mermaid. He specifies that the last of these does not exist in India, although—relying on hearsay—adds that sailors "are reported to have sighted them in adjoining seas".[61] It seems that hearsay continued to supply anecdotes, which appealed to travel writers, even to the self-avowedly sceptical ones such as I'tesamuddin.

It is evident that from Ramusio and Hakluyt (and philosophers like Bacon) onwards, travel writing became aligned with scientific aspirations such as observation, eye witnessing and verification through comparison, a development which was accelerated by the rapid increase in the number of available travelogues. These modifications also guided travel writers to discuss wonders as something that they had seen with their own eyes, as something that was based on a truthful observation.[62] As Daniel Carey has pointed out, travel writers sought to establish their credibility with a number of different strategies, including "simple insistence on their reliability", challenging their doubters' viewpoint and "using testimonials of various kinds".[63]

Chapters of this book

In her chapter "Taming wonder through ekphrasis", María Juliana Gandini discusses the problem of credibility in the context of early modern descriptions of the Spanish Florida. Her analysis of two different treatises—Álvar Núñez Cabeza de Vaca's *La relacion* (1542) and Inca Garcilaso de la Vega's *La Florida del Ynca* (1605)—focuses on the ekphrastic devices in the authors' endeavour to produce reliable descriptions of a "wonder-filled new part of the world". The chapter addresses some of the key problems indicated in this introduction as it demonstrates how lifelike descriptions could make the unknown seem believable. Ekphrastic devices gave words the power to recreate "direct vision", allowing readers to see the wondrous phenomena through their own imagination. In Gandini's treatment of the theme, seeing relies less on visual perception and is achieved by the sheer power of words.

Will Glover's chapter "A country for old men?" explores the wondrous gerontocracy in Walter Ralegh's *The discovery of Guiana* (1596). Starting with Ralegh's fall from Elizabeth's favour in the early 1590s, Glover sets the scene for his interpretation of Ralegh's voyage to Guiana (and the following account of the journey) as an effort to regain the queen's benevolence. Ralegh's first attempt to find "the great and golden citie" of El Dorado with its riches failed, leaving him with the task of seeking for potential investors for his Guiana project by writing a persuasive narrative, which would present the search for New World gold as a feasible endeavour and himself as a reliable narrator. This meant that his writing had to find the right balance between the "wondrous" and the real. Identifying wonders as novel and unexpected phenomena with unknown causes, Glover stresses the importance of "the fundamental *perceived reality*" and excludes supernatural miracles from his study. The wonder that Glover focuses on is the old age of the Indigenous people referred to in Ralegh's treatise. The description of the wondrous gerontocracy in Guiana is analysed as Ralegh's "finely calibrated strategic move".

Mahmoud Khalifa's chapter explores the themes of self-fashioning and the creation of the wondrous in Evliya Çelebi's (1611–82) travelogue Seyahatnâme. Focusing on the Ottoman traveller's account of his journey to Egypt in 1672, Khalifa distinguishes Çelebi's writing from "the paradigm of early modern European self-fashioning", which emphasises antagonism between a "centralized identity" and an "alien" one, a paradigm that aims at reducing and controlling the "threatening Other". In contrast to this, Çelebi's narrative seeks to empathise with otherness. In his analysis of the Seyahatnâme, Khalifa focuses on the grotesque and the carnivalesque, two defamiliarisation techniques that Çelebi employed to criticise the existing social order. The chapter explains Ajaib (the Arabic for wonders) as a fluid notion, not fully detached from either reality or fabrication. What is more, Çelebi "inhabited simultaneously two different worlds, one mundane and the other supernatural". In line with these observations, Çelebi's treatment of the wondrous is ambivalent and, Khalifa concludes, the Other "or the unexplainable" remains somewhere between fact and fiction, losing its threatening unfamiliarity.

Ioana Manea's chapter on François Bernier's (1620–1688) journey to Kashmir focuses on a traveller's attempts to explain wonders by discovering their causes. A physician and a friend of the philosopher Pierre Gassendi, Bernier was influenced by the contemporary "libertine intellectual approach to miracles" and, in line with this, eager to reveal the deception that sustained superstitions. Bernier's travelogue *Voyages* (1670–71) contains an account of his 1664 journey from Delhi to Kashmir as part of the entourage of the Mughal emperor Aurangzeb. In his descriptions of Kashmir—a terrestrial paradise—Bernier uses the term "miracle", which implies a religious quality strong enough to break the natural law. Manea's analysis demonstrates, however, that Bernier's descriptions of wonders were based on rational, not supernatural, explanations of strange phenomena. In his search for the causes of wonders, Bernier resorted to empirical observation, probabilism and the cooperation between reason and imagination.

In her chapter "From strange to familiar: Ottoman eyes in Paris in the eighteenth century", Ipek Bozkaya explores the Ottoman envoy Mehmet Çelebi's journey to Paris in 1721. Her analysis of Çelebi's travelogue *Sefaretname* discusses not only the traveller's experience of the West but also the curiosity and astonishment that Parisians felt in front of the "exotic" Ottoman entourage. The chapter approaches wonder as the tension "which is built around the duality between the observer and the Other" and argues that visual observation—"the capacity of the eye"—plays a key role in the formation of human relations, including the tension that issues from the perceived otherness. While the French were impressed by Çelebi's attire and refined manners (which clashed with old stereotypes), Çelebi was astonished by opera, an art form which he was to introduce to the Ottoman Empire. In Paris, Çelebi was no longer surrounded by familiar habits and ways of thinking. While this filled him with wonder and the sensation of being in another world, Bozkaya concludes that *Sefaretname* demonstrates the "transformative quality of travel and wonder".

The chapter "Wonders and curiosities: early modern British impressions of Bohemia" by Hana Ferencová discusses three travelogues written by the Englishmen Edward Browne (1677), John Swinton (1745–46) and Richard Pococke (1745). What these travelogues have in common—besides Bohemia—is their aspiration towards objective and scientific style and the fact that, despite the scientific bent, they all contain descriptions of novelties, curiosities and extraordinary phenomena. Although Bohemia was located a considerable distance away from the usual Grand Tour route of English travellers, those who diverted from the southern track showed interest in Prague's *Kunstkammer* and in the spas and mines of the region. While the three English travellers discuss spas and curative spring waters strictly from a technical point of view, they address wonders and miracles when describing local legends, folklore and the Catholic veneration of saints and relics. Ferencová points out the English travellers' habit of asserting the reliability of their travelogues by referring to local guides as sources for their descriptions of curiosities.

In his chapter on "wonders travelling from China", Dinu Luca focuses on the collection of rarities accumulated by the Italian scientist Manfredo Settala (1600–80). After a quick overview of Chinese books' presence in Europe and an introduction to Settala's gallery, the chapter focuses on two Chinese texts in the collection. Luca describes Settala as a "maker and a showman", as "the Milanese Archimedes" who amazed his guests not only with creations such as telescopes and automata, but also with the "perhaps 5000-object strong collection" and a library of 10,000 volumes. Luca's detailed analysis of the two Chinese texts reframes previous scholarly conclusions and hypothesises the travels of such material from China to Settala's collection. While "Chinese" was a descriptor on its way to becoming "Europe's favourite short-hand term for 'wonder'", the chapter also points out that "translational tours de force" and "general scholarly disciplining" led to the birth of modern sinology in the early nineteenth century.

The chapter "Reducing and othering wonder" by Antoine Eche explores Abbé Prévost's collection of travel documents *Histoire générale des voyages* (1746–59), a translation and continuation of John Green's *New General Collection of Voyages and Travels* (1745–47). Treating wonder and curiosity as "shifting attitudes" and integral components of the transformations that took place in early modern Western epistemology, Eche analyses eighteenth-century perceptions and evaluations of wonder in the context of knowledge production and demonstrates that wonder played an instrumental role in the European colonial agenda. Travel writing and its employment of wonder as an inescapable trope created a "colonial dynamic", which involved both the traveller and the reader. Rather than "a goad to inquiry", wonder functioned as a "tool for validating bias" and as an "object of investigation" that sustained colonial practices.

In her chapter "Wonders of the night—nocturnal darkness as sensory experience", Susanna Lahtinen discusses travelogues written by four eighteenth- and early nineteenth-century British women, focusing on their experiences of nocturnal travel. Basing her study on insights gleaned from phenomenology and the history of emotions, Lahtinen shows how night time intensified the sense of unfamiliarity and enhanced the feelings of awe and wonder experienced by travellers. For them, darkness was a background that allowed "the moon, stars, flames of the volcanoes, fireflies and the light of the lanterns" to shine. While adorning ruins such as the Colosseum with additional charm, references to the "pale moonbeam" mark some of the travelogues discussed by Lahtinen with an unreserved sense of romanticism. From the point of view of early modern travel, nocturnal darkness had a twofold impact as it made the experience either more challenging or more sublime.

Isabel Gaviria's chapter on "Columbus, travel and wonder" builds an arch from Columbus to twentieth-century Cuban literature, demonstrating how writers such as Lezama Lima and Alejo Carpentier decolonised Columbus's influential depiction of America as a land of "marvel, gold and savages". In the chapter, "marvel" is regarded as an "aesthetic that speaks about otherness" and wonder as a notion that "triggers an aesthetic procedure related to modernity and coloniality". Gaviria notes that "wonder proposes a way of seeing". This is manifested by the fervour with which the European colonisers replaced indigenous gods with Christian imagery, turning the representations of the former into collectable marvels. That Columbus's vision was distorted is fantastically suggested in Carpentier's novel *El arpa y la sombra*, which reveals the emptiness of the marvellous, a concept that needs to be filled with meaning. This final chapter of the book shows how the very notion of wonder travelled, how it experienced a centuries-long decolonising journey filled with epistemic transformations.

Notes

1 Berger 1972, 7.
2 Naipaul 2011, 5, 13–14.
3 Berger 1980, 75.
4 Ivins 1938, 7: "Words as symbols have no meanings except such as they get from general convention or specific agreement as coupled with recognitions arising through concrete experience, and thus are incapable of conveying information about unique characteristics to people who are not acquainted with those characteristics at first hand".
5 Naipaul 2011, 60.
6 Naipaul 2011, 30.
7 Conrad, *The heart of darkness*, 15.
8 Conrad, *The heart of darkness*, 9.
9 Conrad, *The heart of darkness*, 16.
10 For this point, see Rubin 2008, xiii; see also, Said 1994, 32–33.
11 Conrad, *The collected Letters*, vol. 4, 112.
12 Rubin 2008, xi.
13 Conrad, *The heart of darkness*, 32. Said 1994, 24.
14 Pagden 2013, 203.
15 Schaw, *Journal of a lady of quality*, 77, 90. Schaw acknowledges the poor quality of artificial imitation also on page 203, when she refers to Philip Miller's description of magnolias and adds "tho' they are infinitely more beautiful than he describes them".
16 I'tesamuddin, *The Wonders of Vilayet*, 53–54. The reference to "Allah's creative power" correlates with conventional ways of justifying descriptions of the marvellous as manifestations of god's power (for this, see Carey 2019, 544).
17 Said 2019, 4–5; Said 2003, xii-xiii, xxii.
18 Vico, *Principi di scienza nuova*, 138–39: "*ignoranza di cagioni, la qual fu loro madre di maraviglia di tutte le cose... la loro robusta ignoranza... onde furon detti Poeti, che lo stesso in greco suona, che criatori*"; 30: "*vere sentenze poetiche... piene di sublimità, e risveglianti la maraviglia*".
19 Vico, *Principi di scienza nuova*, 140–41: "*Curiosità, ch'è figliuola dell'Ignoranza, e madre della Scienza... volgo, ch'ove veggano o una qualche cometa... o altra stravagante cosa in natura... subito danno nella curiosità, e tutti ansiosi nella ricerca domandono, che quella tal cosa voglia significare*".
20 Vico, *Principi di scienza nuova*, 80: "*È necessario, che vi sia nella Natura delle cose umane una lingua Mentale comune a tutte le Nazioni; la quale uniformemente intenda la sostanza delle cose agibili nell'umana vita socievole, e la spieghi con tante diverse modificazioni per quanti diversi aspetti possan aver' esse cose*".
21 Bryan 1986, 260–61.
22 Alberti, *On painting*, 25, 33.
23 Alberti, *On painting*, 38. Ivins 1938, 13, argues that Renaissance theories on perspective and relativity enabled the subsequent scientific progress and rationality.
24 Alberti, *On painting*, 56.
25 Clark 2007, 1, 123–4. Hakewill, *The Vanitie of the eie*, 57–65.
26 Merleau-Ponty 1968, 11.
27 Merleau-Ponty 1964, 13–14.
28 Merleau-Ponty 1964, 15–16.
29 Boileau, *Traité du sublime*, 9: "*par Sublime, longin n'entend pas ce que les Orateurs appellent le stile sublime: mais cet extraordinaire & ce merveilleux qui frape dans le discours, & qui fait qu'un Ouvrage enleve, ravit, transporte... il faut donc entendre par Sublime dans Longin, l'extraordinaire, le surprenant... le merveilleux dans le discours*".
30 Doran 2007, 356.

31 Longinus, *On the sublime*, 163.
32 Horace, *Ars poetica*, 561, LOEB edition, p. 484: "*ego mira poemata pango*". For the translation, see Francis, *Poetical translation*, 423.
33 Burke, *A philosophical enquiry*, 97–98.
34 Alberti, *On painting*, 59.
35 Schaw, *Journal of a lady of quality*, 91.
36 Purchas, *His pilgrims*, 731.
37 Conrad, *The heart of darkness*, 42, 45.
38 Wollstonecraft, *Letters*, 87.
39 *Oxford English Dictionary*, *s.v.* "wonder".
40 Campbell 1999, 5.
41 Daston and Park 1998, 13, 314. On page 11, Daston and Park quote Francis Bacon's definition of wonder as "the seed of knowledge". See also, Bacon, *Novum organum*, 302–3: "Wonder is a child of rarity" (*Admiratio enim proles est raritatis*).
42 Vico, *Principi di scienza nuova*, 85: "*Il più sublime lavoro della Poesia è, alle cose insensate dare senso, e passione; ed è proprietà de' fanciulli di prender cose inanimate tra mani, e, trastullandosi, favellarsi, come se fussero quelle persone vive*".
43 Conrad, *The heart of darkness*, 57, 91, 95.
44 Some travellers sought to achieve this through study. Carey 2019, 533, mentions a seventeenth-century French diplomat who prepared for his visit to Siam by reading about the region in order to be able to "add light to sincerity and to be well informed to inform others"
45 I'tesamuddin, *The Wonders of Vilayet*, 67.
46 Franklin 2001, 119.
47 Franklin 2001, 119.
48 Negri, *Viaggio settentrionale*, xi–xii: "*non poche cose ho tralasciate… per non aver potuto ottenerne quella certezza, ch'io desiderava… che se bene vere, anno tanto dell'incredibile… che senza colpa fa vergogna*".
49 Rubiés 2024, 36.
50 Acosta, *Natural and Moral History of the Indies*, 8–9.
51 Bacon, *Novum organum*, 133.
52 Bacon, *Novum organum*, 19.
53 Franklin, 123–125. Rubiés 2024, 42, names Columbus, Magellan and Vespucci as "quasi-mythical figures" who, by the end of the sixteenth century, were "celebrated across Europe".
54 Bacon, *Novum organum*, 151.
55 Bacon, *Novum organum*, 137.
56 Bacon, *Novum organum*, 87.
57 Franklin 2001, 125.
58 Rubiés 2024, 38–40.
59 Rubiés 2024, 44–5. See page 51 for further discussion on the use of various sources to establish the veracity of a travelogue.
60 I'tesamuddin, *The Wonders of Vilayet*, 29. The same point is made in Negri, *Viaggio settentrionale*, xi, where he quotes St. Jerome: "*aliter visa, aliter audita narrantur, quod melius intelligimus, melius proferimus*" (things that we have seen are told in one way and things that we have heard in another, the better we understand things, the better we describe them).
61 I'tesamuddin, *The Wonders of Vilayet*, 28, 46–48.
62 Rubiés 2024, 62–3. According to Rubiés, in this period (sixteenth century) "the marvellous" became "increasingly understood as the new, the extraordinary or the monstrous within nature, rather than the miraculous".
63 Carey 2019, 526.

Bibliography

Primary sources

Acosta, De José. *Natural and Moral History of the Indies*. Edited by Jane E. Mangan, translated by Frances López-Morillas. Duke University Press, 2002.

Alberti, Leon Battista. *On painting*. Edited by Rocco Sinisgalli. Cambridge University Press, 2011.

Bacon, Francis. *Instauratio magna*, Part II: *Novum organum and Associated Texts*. Edited and translated by Graham Rees with Maria Wakely. Oxford University Press, 2009.

Boileau, Nicholas. Traité du sublime, ou du merveilleux dans le discours. In *Boileau, Nicholas, Oeuvres diverses*, vol. II. Amsterdam, 1707.

Burke, Edmund. *A philosophical enquiry into the origin of our ideas of the sublime and beautiful*. London, 1768.

Conrad, Joseph. *The collected Letters*, vol. 4, 1908–1911. Edited by Frederick R. Karl and Laurence Davies. Cambridge University Press, 1990.

Conrad, Joseph. *The heart of darkness*. Everyman, 1995.

Francis, Philip. *Poetical translation of the works of Horace*. Edinburgh, 1779.

Hakewill, George. *The Vanitie of the eie*. Oxford, 1608.

Horace. *Ars poetica*. Translated by H. Rushton Fairclough. LOEB, Harvard University Press, 1929.

I Tesamuddin Mirza Seikh. *The Wonders of Vilayet, Being the Memoir, originally in Persian, of a Visit to France and Britain in 1765*. Translated by Kaiser Haq, Peepal Tree Press Ltd: Leeds, 2002.

Longinus. *On the sublime*, translated by W.H. Fyfe. LOEB, Harvard University Press, 1995.

Negri, Francesco. *Viaggio settentrionale*. Forli 1701.

Purchas, Samuel. *His pilgrimes in five bookes, the first part*. London, 1625.

Schaw, Janet. *Journal of a lady of quality*. Yale University Press, 1923.

Vico, Giambattista. *Principi di scienza nuova*, vol. I. Naples, 1744.

Wollstonecraft, Mary. Letters written during a short residence in Sweden, Norway, and Denmark (1796). In *Travel Writing 1700-1830: An Anthology*, 82–95. Edited by Elizabeth A. Bohls and Ian Duncan. Oxford University Press, 2005.

Literature

Berger, John. *About looking*. Bloomsbury Publishing, 1980.

Berger, John. *Ways of Seeing*, Penguin Books, 1972.

Bryan, Ferald J. "Vico on metaphor: Implications for rhetorical criticism". *Philosophy & Rhetoric* 19, no. 4 (1986): 255–266.

Campbell, Mary Baine. *Wonder & science: imagining worlds in early modern Europe*. Cornell university Press, 1999.

Carey, Daniel. "The problem of credibility in early modern travel", *Renaissance Studies* 33, no. 4 (2019): 524–547.

Clark, Stuart. *Vanities of the Eye: Vision in Early Modern European Culture*. Oxford University Press, 2007.

Doran, Robert. "Literary history and the sublime in Erich Auerbach's 'Mimesis'". *New Literary History*, vol. 38, no. 2 (2007): 353–369.

Franklin, Wayne. "Language and event in New World history". In *Defining Travel: Diverse Visions*, edited by Susan L. Roberson, 117–131. University Press of Mississippi, 2001.
Ivins, William M. "On the rationalization of sight with an examination of three Renaissance texts on perspective". *Papers*, no. 8, The Metropolitan Museum of Art, New York, 1938.
Merleau-Ponty, Maurice. "The Primacy of Perception and Its Philosophical Consequences". Translated by James M. Edie. In *Merleau-Ponty, Maurice, The Primacy of Perception*, edited by James M. Edie, 12–42. Northwestern University Press, 1964.
Merleau-Ponty, Maurice. *The Visible and the Invisible*. Translated by Alphonso Lingis. Northwestern University Press, 1968.
Naipaul, V.S. *The enigma of arrival*. Picador, 2011.
Pagden, Anthony. *The Enlightenment and why it still matters*. Random House, 2013.
Rubiés, Joan-Pau. "Distance and credibility in sixteenth-century travel writing: Discovery, text, and truth in Varthema, Vespucci, and Pigafetta". In *Far From the Truth: Distance, Information, and Credibility in the Early Modern World*. Edited by Michiel van Groesen and Johannes Müller, 36–68. Routledge, 2024.
Rubin, Andrew N. "Foreword". In Said, Edward W, *Joseph Conrad and the fiction of autobiography*, ix–xviii. Columbia University Press, 2008.
Said, Edward. *Culture & Imperialism*. Vintage Books, 1994.
Said, Edward. "Introduction". In Auerbach, Eric, *Mimesis: The representation of reality in western literature*, ix–xxxii. Princeton University Press, 2003.
Said, Edward. *Orientalism*. Penguin Books, 2019.

2

TAMING WONDER THROUGH EKPHRASIS

Florida and Europe, 1542 and 1605

María Juliana Gandini

Making sense out of wonder

Constructing the so-called New World as an object of knowledge and conquest was a complex task for early modern Europeans. First, it was difficult to understand, and at the end to convince others of the veracity of the information collected about new and strange lands, peoples and natural phenomena, often described as nothing short of "marvellous".[1] Second, during the sixteenth and early seventeenth centuries, knowledge about the world was supposed to be bound—to a great extent—to the works of ancient and biblical authorities.[2] Third, even if the accounts of America's first-hand witnesses were to be believed by their European audiences, their discrepancies about the characterization of the New World and its inhabitants led to suspicion or even distrust.[3]

In this context, truth and trust building were delicate, polemic issues. Explorers and colonists had to find words and ideas to portray geographies, animals, plants and peoples far removed from their collective cultural traditions. Moreover, the governance of the colonies needed reliable information, but governments and their agents were overwhelmed by "mountains of papers with outcries"—as Pablo Neruda aptly wrote—that sometimes portrayed opposite versions of the same events. Questions about what to believe thus arose: What to do with the information about wondrous, out-of-the-ordinary episodes, such as native peoples' strange customs or tales of improbable survival on unfamiliar lands? How to deal with the uncertainty and distortion entailed by physical distance? Who was lying, who was telling the truth?

To build compelling, trustworthy accounts about a wonder-filled new part of the world—accounts that were aimed not just to be believed but to be

DOI: 10.4324/9781003506928-2

regarded as more veracious than others—its witnesses took advantage of every strategy at hand to achieve that goal.[4] They could appeal to traditional means, such as the reference to ancient, authoritative authors, or persuasion techniques found in classic rhetoric. They could also employ more novel tools like reinforcing their own quality as eyewitnesses over unprecedented experiences or channel what Roger Chartier called "the powers of print".[5] Finally, references to witnesses' faith, social status, benefactors or institutional affiliations and duties were routinely used and abused as ways to sustain authority.[6] In any case, the description of the New World and the verisimilitude of the representations produced about it were rooted in a culture of truth that mingled different elements: bookish erudition, rhetoric devices, direct observation, validation of empirical data and social conditions.

Phenomena referred to as marvellous constitute a prolific topic with which we can analyse the strategies deployed to construct verisimilitude over what was characterized as new, unknown and surprising to early modern European reading publics. In a more general sense, Lorraine Daston and Katharine Park studied how wonders and wonder drew the frontiers between what was considered natural, known, domestic and cultivated on the one hand and what was regarded as unnatural, unknown, exotic and vulgar on the other.[7] Considering travel narratives, François Hartog highlighted how the inclusion of wonders' descriptions in them produced an effect of trustworthiness by itself: finding admirable, wondrous differences while visiting foreign lands and peoples was something to be expected and contributed to the credibility of the witness-traveller.[8] In this way, wonder and credibility reinforced each other in the context of European overseas expansion: as the former questions the range of credibility, the latter "insists upon the undeniability, the exigency of the experience".[9] This link was exploited by those who intended to construct themselves as authorities over the New World.

In this setting, I will analyse two founding narratives of Spanish Florida: Álvar Núñez Cabeza de Vaca's *La relacion* (1542) and Inca Garcilaso de la Vega's *La Florida del Ynca* (1605). Cabeza de Vaca was a—quite unfortunate—first-hand witness of Florida who wrote a famous account of his experiences between 1527 and 1536 as a defeated conqueror, shipwreck survivor, captive and shamanic healer. Instead, Garcilaso wrote about other people's ordeals in Florida, producing a history book based on written and oral accounts of several survivors of Hernando de Soto's expedition (1539–1543). Both authors had to construct themselves as authorities over Florida, using different means to do so according to their diverse social positions and objectives. Cabeza de Vaca was a witness-protagonist of the events recollected in his text, procuring to benefit his career after the disastrous failure of Pánfilo Narváez's expedition in 1527–1528. Garcilaso, with his unstable standing as *mestizo*, aimed to be recognized as a competent historiographer, able to articulate and elevate diverse testimonies in an overarching narrative to support his claims about

American peoples' dignity. Both authors needed to convince their audiences of the truthfulness of their accounts. I will focus on one particular strategy used in *La relacion* and in *La Florida* to convey vivid descriptions of what is presented as wondrous events: the ekphrasis.

Ekphrasis has an ancient ancestry, which could be traced back to the critical description of Achilles' second shield included in the *Iliad*.[10] Later on, no classical orator, lawyer, geographer or historian could ignore it, for ekphrasis provided the means to present vivid descriptions of events, objects, natural phenomena, persons or places to their audiences. In these settings, ekphrastic devices appealed to convince through the move of affections and the construction of rational proof.[11] Finally, ekphrasis could render lifelike descriptions of works of art, a function which has attracted the attention of modern scholars but was by no means dominant in its long history.[12]

Considering these broader subjects, it is easy to see how ekphrasis was an important strategy for building knowledge about the world in the context of European overseas expansion.[13] In particular, I will analyse how both Cabeza de Vaca and Garcilaso invoked the power of ekphrasis to communicate and convince their readers of the wonderworthy capabilities of native peoples' warfare. The detailed description of the exploits of the latter aimed to elicit admiration and surprise from the European readers, presenting the combat developments as wonders. At the same time, ekphrastic devices explained Floridian war costumes overcoming the idea of native peoples' supposed weakness. They helped to establish Florida's Indians as worthy enemies, both picturing their feats as directly seen and mimicking the surprise—translated as wonder—of the European readers as they imagined them.

Cabeza de Vaca and Garcilaso tamed wonder through ekphrasis, making it a tool to produce authority, knowledge and plausible explanations over the New World. In a broader perspective, the analysis of their verisimilitude construction devices illustrates the constant necessity for what Juan Pimentel identifies as "*powerful words*, accurate metaphors, and careful plot strategies" in the context of the early modern overseas expansion, when distance and credibility found themselves usually at odds.[14]

Enárgeia, verisimilitude and historic truth

Since the European audiences expected authors to prove the credibility of their accounts about the New World, it is important to study the mechanisms that constructed verisimilitude.[15] Although ekphrasis is not usually included in the genealogy of modern empiric knowledge, the rhetorical possibilities present in lifelike descriptions played a key role in making believable a part of the world unknown to ancient authorities.

Ekphrastic devices relied on *enárgeia*, the power of words to recreate—or to play as—direct vision of any given phenomena, making the reader or the audience see it through their imagination.[16] This power of words aimed to stir readers' emotions and sympathies towards the ends desired by the author. However, the management of affections was not the only potential of ekphrasis, as it also related to rational proof construction through the artificial recreation of *autopsia*, direct visual perception.[17]

Early modern historiographers relied on ekphrastic devices to provide the illusion of *autopsia*, assembling a historical text as a true account. Beyond the Aristotelian principle of history as a genre obliged by truth, historians appealed to the power of ekphrasis to endow with vivacity their descriptions of past events, persons or places. This quality of ekphrasis as a verisimilitude-creating device led John Slater to assess early modern Spanish historiography as an ekphrastic genre, in which historical authority was constructed as a visual effect.[18] Although Slater is primarily concerned with Peninsular historiography and its relationship with ekphrasis about works of art, he briefly acknowledges the importance of eyewitness testimony modelled after judiciary procedure as a strategy to build reliability for their accounts.[19] It is not difficult to see how these two aspects are not only close but deeply interrelated. First-hand witnesses were also using ekphrastic devices to present lifelike, trustworthy descriptions of what they observed and experienced, and historians—just like lawyers and judges—relied on such capacity to evaluate the verisimilitude of testimonies.

But how did ekphrasis accomplish this goal? The detailed description entailed in ekphrastic devices created the illusion of immediate vision for their audiences even when they were clearly removed—in different degrees—from the original phenomena that were being described. The accounts could also include, besides the representation of the events themselves, the point of view and reactions of those who allegedly had direct vision over the incidents in question. In this way, ekphrastic devices could tackle suspicion or doubts about credibility pointing out how first-hand witnesses were also moved, and perhaps even a little sceptical, about what was—supposedly—happening in front of their eyes.

Both Álvar Núñez Cabeza de Vaca and Garcilaso de la Vega leveraged the evocative powers of ekphrasis to make credible accounts about episodes that they suspected could cause disbelief among their readers. The combination of direct perception, lush details and witnesses' reactions provided writers with important internal textual tools to create verisimilitude: Cabeza de Vaca constructed himself as a witness-protagonist whereas Garcilaso presented his own witnesses as reliable individuals, and his account as a truthful rendition of their testimonies. I will explore the strategies deployed by both authors in their descriptions of wonder-like events related to Indian warfare in today's Florida peninsula. The attention devoted to native peoples' war practices was a crucial element in Cabeza de Vaca's and Garcilaso's ambitious projects. The former had to justify how he ended up stranded in Florida, while the latter was invested in establishing the dignity and rational qualities of American natives.

Experience and authority in Florida's description

Álvar Núñez Cabeza de Vaca passed by Florida in 1528, serving as treasurer of Pánfilo de Narváez's expedition. His landing in the area was the beginning of what he later defined as a "so sad and wretched a captivity".[20] The expedition was rapidly thrown into disarray while exploring Florida, and later it was dissolved following a risky sea voyage that ended tragically on Galveston Bay. Soon after, only Cabeza de Vaca and three or four other survivors were alive, stranded in present-day Southwest United States and Northern Mexico for almost a decade.

In 1536, Cabeza de Vaca and his three voyage companions returned to Spanish-populated areas in Mexico. He started to write an account of what he had experienced in the New World, hoping that, together with his family's influence, the report could gain him the king's favour and some kind of recognition, such as a new position in his service. The famous proem of *La relacion* states this intention using the language of service to the king, which enabled Cabeza de Vaca to transform a categorical failure into a tribute of information to the sovereign about Florida, a tribute made possible by the power of the quill:

> I had no opportunity to perform greater service than this, which is to bring to Your Majesty an account of all that I was able to observe and learn in the nine years that I walked lost and naked through many and very strange lands.[21]

Cabeza de Vaca's authority was supported by his role as a direct eyewitness to the experiences referred to in his text. This first-hand knowledge, made possible through extraordinary labours, was the key element that allowed him to construct himself as an author. It also helped him to authorize his account's truthfulness without having the traditional social and educational credentials of a cosmographer or a historian.[22]

Garcilaso de la Vega could not claim that kind of direct knowledge about Florida as he did later in his books dedicated to Peru. Instead, he asserted, as far as he could, his "Indian" condition to explain several characteristics of Florida's native peoples in relation to his first-hand expertise of the Andean region. But that was not enough: his authority as a historian of the Hernando de Soto expedition had to be established on different grounds. He was not a newcomer in erudite matters, as he distinguished himself as a skilful translator of Judah Leon Abravanel's *Dialogues of Love* into Spanish.[23] But he still had to become a historian, presenting his account as a carefully crafted compilation of trusted eyewitness' testimonies: in his modest—and deceitful—words, he claimed that as an author, he "contributed no more than the pen".[24] Garcilaso stated that *La Florida del Inca's* main source was the oral account of a captain whom he referred to as "his author", and who has been identified as Gonzalo Silvestre.[25] They met in Peru some years after the latter's survival from the disastrous De Soto expedition and reencountered afterwards in Spain. Besides this direct and privileged testimony, Garcilaso claimed to have considered two short *relaciones*—accounts or reports—written by Alonso de Carmona and Juan de Coles, who also took part in the expedition.[26] Garcilaso's history of the failed conquest of Florida is thus presented as a fusion of direct testimonies of other people, articulated and elevated to a historiographical level by his authorial intervention.

Using a personal, eyewitness testimony or otherwise working on someone else's, Álvar Núñez Cabeza de Vaca and Garcilaso de la Vega supported the credibility of their works with rhetorical devices. Ekphrasis could translate visual perceptions into narrative forms and, in so doing, create plausible accounts. In this way, the old illusion of direct vision was recreated through words and put at the service of making the wonders of the New World credible.

The wonders of Florida's warfare made credible through ekphrasis

Cabeza de Vaca didn't write much about the passage of Pánfilo de Narváez's expedition through Florida in 1528. In the 1555 Valladolid edition, this part of the journey is contained in seven of the thirty-eight chapters that compose the book.[27] The narrative is focused on the reasons for the expedition's disintegration: on the one hand, Narváez's ineptitude; on the other, the continued harassment of the natives defending their land, which ultimately forced the invaders to abandon any conquest project and to leave Florida in five precarious rafts.

The direct eyewitness perspective is evident when Cabeza de Vaca refers to the death and dangers imposed on the aspiring conquerors by Floridian natives. Although their main weapons were bows and arrows, their warfare practices were quickly taken as a threat and described as wondrously consequential. Writing about a confrontation over the passage of a lagoon, Cabeza de Vaca mingles the surprise caused by the lethality of the arrow shots and the horror provoked by the possibility of being one of their victims:

> In this struggle there were some of our wounded who could not make use of the good weapons they carried, and there were men this day who swore that they had seen two oaks, each one of them as thick as a man's lower leg, pierced through and through by the arrows of the Indians, and this is not to be wondered at, having seen their strength and skill with which they shoot them, because I myself saw an arrow in the base of a poplar tree that had entered it to the depth of a *xeme*.[28]

In this passage, the ekphrastic device is deployed with unambiguous measurements: the pierced oaks were as thick as a man's leg; the arrows entered a poplar a *xeme* deep (the distance between the point of the thumb and that of the index finger, both extended in different directions). The description also states who observed these prodigies. In the first case, several men referred to the event, so the verisimilitude of the testimony was thus supported; in the second one, the observer was Cabeza de Vaca himself, the author of the account who had sworn in the king's name that he

> wrote all this with such sure knowledge that although some very novel things may be read in it, very difficult for some to believe, they can absolutely give them credence.[29]

The power of the arrow shots, expressed through measurements and validated by direct witnesses, demonstrated how severely the Spanish were wounded when confronting Florida's natives. Ekphrastic description of what was seen enabled the representation of the marvellous efficacy of simple weapons to a Spanish audience. At the same time, ekphrastic devices could explain that the power of the arrows was not something truly marvellous, removed from rational comprehension. Cabeza de Vaca tamed their seemingly wondrous qualities by introducing a detailed description of the skill and characteristics of the native archers:

> All the Indians we had seen from Florida to here are archers, and as they are of large build and go about naked, from a distance they appear to be giants. They are people wonderfully well built, very lean and of great strength and agility. The bows they use are as thick as an arm [and] eleven or twelve spans long so that they can shoot arrows at two hundred paces with such great skill that they never miss their target.[30]

These claims were explained again through accurate measurements: the bows were enormous, so it was logical to think that the people who used them were physically strong and thus able to perform the prodigies stated above. But the explanation did not completely erase the marvel of the natives' warfare: it endured as admiration—they were indeed superb archers with athletic bodies—and as an illusion, as they looked like giants from afar.

In other instances, Cabeza de Vaca used ekphrastic devices to explain the harm done to the human body. He describes how the Floridians killed a *hidalgo* called Avellaneda, striking "him with an arrow at the edge of his breastplate, and the wound was such that almost the entire arrow passed through his neck, and later he died there."[31] A few pages further on, he writes that the hungry conquerors were collecting shellfish in a marsh when the natives

> killed ten of our men within sight of camp without our being able to rescue them; we found them shot through and through with arrows, and although some of them had good weapons, their arms were not sufficient to prevent this from happening because the Indians shot arrows with as much skill and force as I have described above.[32]

In both cases, ekphrasis organizes the account, linking the direct sight of the eyewitness with the details of what was observed. In the latter case, there were multiple witnesses—including the author—who were able to describe the events as something they had seen with their own eyes. The account duplicates the story about the arrow-pierced oaks, although this time the target was not wood but human flesh. The event took place in the vicinity of the Spanish camp. However, the conquistadors failed to notice the presence of the natives—who had been observing them—until it was too late, and ten Spanish were dead. The stealth of the natives was yet another wonder to perplex the Spanish, although noticeable only by its horrifying results.

Garcilaso de la Vega's arsenal of rhetorical resources was more sophisticated than the one deployed by Álvar Núñez Cabeza de Vaca. That said, Garcilaso was a keen reader of *La relacion*.[33] Surpassing what his informants could have told him about Florida in the context of De Soto's expedition, Garcilaso acknowledged his debt to Cabeza de Vaca's account, a debt easily recognizable on the topic of warfare.

Garcilaso dedicated fifty-five chapters (the first two parts of *La Florida's* Book Two) to the progress of De Soto's expedition during 1539 in the same territories explored by Narváez in 1528. The extent of *La Florida* produces a large number of detailed descriptions of military actions performed by Florida's native peoples. But besides the mere number, we must consider one of Garcilaso's key objectives in writing about the history of the conquest of America: to contest what he defined as the Europeans' ordinary opinion of the American Indians as "simple folk without reason or understanding who in both peace and war differ very little from beasts."[34] Since his goal was to defend the dignity of American natives, it is easy to understand why *La Florida* brims with detailed accounts of their bravery and eloquence, stressing their ability to defend themselves from the conquerors.

In *La Florida*, the descriptions of these capacities through ekphrastic devices are clearly constructed, referring not just to the battles and confrontations but to the reactions of those Spanish conquerors who took part in them. The marvellous dimension attributed to the events is linked to the textual representations of their surprise, wonder or even terror, emotions that are carefully staged to duplicate those imagined in the readers.

Take, for example, the descriptions involving the fearsome and cunning chief Vitacucho of Apalache. Forced to endure the occupation of his village by De Soto and his men, he initiated a planned revolt, attacking the Spanish during their dinner. De Soto only survived because of the intervention of his companions, who sprang to action at the sight of a "strange and unexpected performance"[35]:

> Seizing him [De Soto] by the collar with his left hand, he [Vitacucho] gave him such a blow over eyes, mouth and nose with his right fist that he knocked down the chair in which he was seated and stretched him out unconscious in his back as if he had been a child. Then in order to finish with his victim, he let himself fall upon him, while at the same time giving such tremendous roar that it could be heard for a quarter of a league around.[36]

Then, an equally comprehensive description of the effects of Vitacucho's assault on De Soto was included:

> he was unconscious for more than a half hour, and he bled through the eyes, nose, mouth, gums, and upper and lower lips as if he had been struck with a large club. Furthermore both his front and back teeth were so tormented that they almost fell out, and for upwards of twenty days he could eat only food that could be taken with a spoon and nothing that had to be chewed. Indeed during this whole time his nose and lips remained so swollen that it was necessary to apply plasters.[37]

The consequences were as wondrous as the attack itself. To construct it as such, Garcilaso deployed a visual, anatomically oriented ekphrasis of the state of De Soto's face: he mentions one by one the injured areas (eyes, nose, mouth, gums, lips), notes his sore teeth and ends the description of the conquistador eating soft food with a spoon like a child. The full description of Vitacucho's assault highlights his violent, unconscionable character, which matches his formidable physical presence. Although this portrait is not favourable, it does not invalidate the exemplary and positive qualities attributed to Floridian Indians as a collective, considering the rich gallery of otherwise noble and exemplary natives described by Garcilaso.[38] Furthermore, it is possible to think that the author compensated the unflattering description of Vitacucho in two other ways: first, by the implicit fact that he was defending his people; second, because Garcilaso modelled the fierce disposition of the chief after the highly regarded chivalric characters of Matteo Maria Boiardo and Lodovico Ariosto's epic poems.[39]

What is undeniable is the visual performance put together by Garcilaso to present the wondrous physical vigour of American natives, portrayed in the description of Vitacucho's attack on De Soto and in the ekphrasis of the consequences of the assault. The verisimilitude of the account is supported by the specificity of the description, by the affirmation that the story was a recollection based on the testimonies of trusted eyewitnesses and by the depiction of the admiration and surprise felt by those who were present during the incident. Garcilaso makes his readers see the fateful dinner at Vitacucho's town as if they were the astonished companions of De Soto watching him being plummeted by strikes in the middle of what was supposed to be a peaceful meal.

Like Cabeza de Vaca's account, Garcilaso's text contains several portrayals of the efficiency of Florida's archers and of the attacks inflicted on De Soto's army, descriptions imbued with references to wonder. Following three steps, Garcilaso builds carefully crafted descriptions of the power of the archers' arrows. First, he presents a detailed account of the confrontation during which the shots were fired. Next, he portrays the surprise that the Spanish felt because of the archers' strength, accuracy and deadliness. Finally, Garcilaso explains how the conquerors sought to understand the danger imposed by the natives' weapons by conducting different kinds of experiments. These phases form a chain which starts with facts, moves into a description of wonder and ends with a rational explanation of what seemed wondrous at first sight. All three stages are presented through ekphrastic devices, which rely on strong details and on the reconstruction of the eyewitness reactions.

Following these steps, Garcilaso describes a confrontation between a mounted Spaniard and an Indian archer. The latter, seeing that the horseman was soon to reach him "turned to receive his pursuer with an arrow placed in his bow".[40] Even when the Indian died by the spear of the rider, he was able to kill his horse. After this description, Garcilaso commented on the witnesses' impressions: they were "astonished that an animal so gallant, hardy and spirited as a horse should have been slain so suddenly by a single arrow fired at such close range".[41] This wondrous event was investigated further the next morning when the Spaniards examined the dead horse. The findings were described using the power of ekphrasis, allowing Garcilaso to create—or to quote, if we take his verisimilitude protests seriously—the direct vision of the eyewitnesses, the *autopsia*:

> On opening the horse, they found that after entering the chest, the arrow had passed through the center of the heart, the stomach and the intestines, and finally had stopped in the bowels.[42]

The power of the arrow's shot was thus "shown" to the readers of *La Florida* through the harm perceived in the horse's body, listing every injured organ as proof of the native peoples' bravery, strength and skill.

In a discussion not dissimilar to that in Cabeza de Vaca's work, Garcilaso reveals that the wondrous qualities of the archers were the result of continuous practice, which the natives started already in their early years. His description of the practice of archery among Floridian natives is vibrant and lifelike, indicating how infants "ask for bows and arrows" and if "denied these things, they themselves make them".[43] Furthermore, the young natives used their weapons to hunt mice and lizards during "three, four and six hours with the greatest imaginable attention and with their arrows set in their bows", and when they could not find this kind of mark, they hunted flies.[44]

These observations on the practice of archery strengthened the credibility of the episodes described in *La Florida*. In one such episode, the victim was the *maestro de campo* Luys de Moscoso. His companions were "amazed" (*admirados*) by a shot that passed through an expensive armour and pierced "a leather jacket and the hauberk which he wore underneath as well".[45] The episode inspired the Spaniards to test their armours against the Indian arrows. They commanded an enslaved Appalachian to perform a demonstration for them, making a target out of a basket covered with one of their coats of mail. His first shot "penetrated both hauberk and basket so freely and violently that it would have gone on through a man had there been one on the other side".[46] A second test was required, this time adding a second coat to the target. The conquerors even taunted the archer, asking "if he were man enough to pass the missile through" both.[47] The result was mortifying: the arrow passed the coats "at the center of the basket and after passing through the four thicknesses of mail, remained crossed therein with an equal portion of the arrow protruding on each side".[48] The Appalachian man wanted to perform a third shot, hoping to completely clear the second mark and betting his life on it. Unwilling to be humiliated again, the conquerors refused his petition. From this point onwards, the Spanish started to replace their ineffective coats of mail (which they now referred to as delicate "Flemish linens") with Floridian textiles.[49]

These vivid descriptions display apparent wonders as part of Floridians' ordinary life. What is initially presented as something wondrous or miraculous is gradually provided with a solid, rational basis, making the phenomenon understandable and believable. Ekphrastic devices enhanced the verisimilitude and credibility of Garcilaso's account, representing local warfare practices with vivid details and manipulating the reactions of his readers: the descriptions of surprise, admiration and fear experienced by the eyewitnesses were expected to be duplicated by his audience. Garcilaso's readers were guided in consecutive steps—from wonder to rational proof—with the help of different exercises of ekphrasis. The readers had little chance to question the credibility of the events, which—albeit through the vivid words of others—were presented as unfolding in front of their (surrogate) eyes.

Some conclusions on the power to make see as a way to domesticate wonder

Although ekphrasis served Cabeza de Vaca and Garcilaso's objectives, it had its limits. The characteristics of the New World alongside the travails experienced during its conquest were sometimes indescribable. "I refrain here from telling this at greater length because each one can imagine for himself what could happen in a land so strange", Cabeza de Vaca remarked just before describing how Pánfilo de Narváez's army abandoned Florida.[50] In a similar vein, Garcilaso stated that he lacked "sufficient words to present in their proper light the actual truths that are offered me in this history".[51] These were rhetorical strategies that combined false modesty with an appeal to readers' imagination. In contrast to ekphrasis, these strategies do not portray a picture built by words but one built by the author's silences and suggestions.

These elisions are symptomatic of the difficulties of explaining the American world and making oneself a reliable authority over it. Cabeza de Vaca and Garcilaso's readers surely harboured doubts about the truthfulness of both authors. There was no shortage of accounts, histories and cosmographies capable of contradicting or at least disputing their narratives. In the end, Cabeza de Vaca, Garcilaso and a plethora of authors who engaged in the description of the Americas had to deal with the same challenge of representing and constructing authority over a far-away world unknown to European traditions. In this context, any strategy that could establish and sustain authority was deemed helpful. While the standards of observation and data validation were not yet set, old-fashioned methods to produce truthlike representations could be successfully deployed over new subjects, helping shape the norms of verifiable reporting along the process. The ancient technique of ekphrasis was called to aid in constructing credible accounts about the New World and in rationally explaining its characteristics; in translating visual phenomena into a narrative form and in turning vivacity into verisimilitude. In the hands of the authors of such accounts, ekphrastic devices showed and suggested, convened astonishment, offered explanations and charmed readers into being witnesses by proxy. For these reasons, ekphrasis played a crucial role in the slow transformation of the New World into a tamed, cognizable object.

Notes

1 Greenblatt 1991, 14, 19–20.
2 Grafton 1992, 58.
3 Pagden 1993, 57–58, 81–83.
4 Carey 2019, 524–47.
5 Chartier 2010.
6 González Echevarría 1990, 59–60; Shapin 1994.
7 Daston and Park 1998, 20.
8 Hartog 1988, 237.
9 Greenblatt 1991, 20.
10 Becker 1995; Francis 2009, 1–23.
11 D'Angelo, 1998, 439–47.
12 Bartsch and Elsner 2007, i–vi; Elsner 2010, 10–27.
13 Cummins 1995, 152–74.
14 Pimentel 2003, 70. See also, Van Groesen and Müller 2023, 1–18.
15 Myers 1990, 616–25; Voigt 2009, 40–98; Kwiatkowski 2017, 43–94.
16 Preminger and Brogan 1993, 332; Ginzburg 2012, 8–10.

17 Ginzburg 2012, 11–12. For *autopsia* in the description of American phenomena, see Pagden 1993, 51–87.
18 Slater 2007, 217–32.
19 Slater 2007, 220.
20 Núñez Cabeza de Vaca, "La Relación", 244–45: "*tan triste y miserable captiverio*".
21 Núñez Cabeza de Vaca, "La Relación", 18–19: "*no me quedó lugar para hazer más servicio deste, que es traer a Vuestra Magestad relación de lo que en nueve años por muchas y muy estrañas tierras que anduve perdido y en cueros, pudiesse saber y ver.*"
22 For the traditional credentials of a royal chronicler, see Kagan 2009, 72; for other strategies in the writing of history of the conquest, see Adorno 1997, 156–58; Bauer 2003, 35–36.
23 Garcilaso de la Vega, *La Traduzion del Indio de los Tres Dialogos de Amor de Leon Hebreo*, Madrid 1590.
24 Garcilaso de la Vega, *The Florida of the Inca*, 158; Garcilaso de la Vega, *La Florida del Ynca*, f. 82v: "*yo no puse mas de la pluma como escriuiente*".
25 Miró Quesada 1994, 171–72.
26 Mora Valcárcel 2006, 215–16.
27 Núñez Cabeza de Vaca, *La relacion y comentarios*, Valladolid 1555.
28 Núñez Cabeza de Vaca, "La Relación", 62–63:

> En esta rebuelta huvo algunos de los nuestros heridos que no les valieron buenas armas que llevavan, y huvo hombres este día que juraron que avían visto dos robles, cada uno dellos tan gruesso como la pierna por baxo, passados de parte a parte de las flechas de los indios, y esto no es tanto de maravillar vista la fuerça y maña con que las echan, porque yo mismo vi una flecha en un pie de un álamo que entrava por él un xeme.

29 Núñez Cabeza de Vaca, "La Relación", 20–21: "*escreví con tanta çertinidad que aunque en ella* (*La Relación*) *se lean algunas cosas muy nuevas y para algunos muy difficiles de creer, pueden sin dubda creellas*".
30 Núñez Cabeza de Vaca, "La Relación", 62–63:

> Quantos indios vimos desde Florida aquí todos son flecheros, y como son crescidos de cuerpo y andan desnudos, desde lexos paresçen gigantes. Es gente a maravilla bien dispuesta, muy enxutos y de muy grandes fuerças y ligereza. Los arcos que usan son gruessos como el braço de onze o doze palmos de largo que flechan a dozientos passos con tan gran tiento que ninguna cosa yerran.

31 Núñez Cabeza de Vaca, "La Relación", 64–65: "*con una flecha por el canto de las coraças, y fue tal la herida que passó casi toda la flecha por el pescueço y luego allí murio*".
32 Núñez Cabeza de Vaca, "La Relación", 72–73:

> los indios en dos vezes que dieron en ellos nos mataron diez hombres a vista del real, sin que los pudiéssemos socorrer los quales hallamos de parte a parte passados con flechas que, aunque algunos tenían buenas armas, no bastaron a resistir para que esto no se hiziesse, por flechar con tanta destreza y fuerça como arriba he dicho.

33 Adorno 2006, 149–79.
34 Garcilaso de la Vega, *The Florida of the Inca*, 157; Garcilaso de la Vega, *La Florida del Ynca*, f. 82r-82v: "*son tenidos por gente simple sin razon ni entendimiento, y que en paz y en guerra se han poco mas que bestias*".
35 Garcilaso de la Vega, *The Florida of the Inca*, 164; Garcilaso de la Vega, *La Florida del Ynca*, f. 85v: "*un hecho tan estraño y nunca imaginado*".

36 Garcilaso de la Vega, *The Florida of the Inca*, 163–64; Garcilaso de la Vega, *La Florida del Ynca*, f. 85v:

> asiendole con la mano yzquierda por los cabeçones, con la derecha apuño cerrado le dio vn tan gran golpe sobre los ojos, narizes y boca, que sin sentido alguno, como si fuera vn niño, lo tendio de espaldas a el y a la silla en que estaba sentado: y para acabarlo de matar, se dexo caer sobre el, dando vn bramido tan rezio, que vn quarto de legua encontorno se pudiera oyr.

37 Garcilaso de la Vega, *The Florida of the Inca*, 164; Garcilaso de la Vega, *La Florida del Ynca*, f. 86r:

> estuvo el Gobernador más de media hora sin bolver en si, y le hizo rebentar la sangre por los ojos, narizes, boca, enzias y labios altos y baxos, como si le dieran con vna gran maça. Los dientes y muelas quedaron de tal manera atormentados, que se le andauan para caer, y en mas de veynte dias no pudo comer cosa que se vubiese de mascar, sino viandas de cuchara. El rostro particularmente las narizes y labios quedaron tan hinchados que en los veynte dias vuo bien que emplastar en ellos.

38 Garcilaso de la Vega, *La Florida del Ynca*, f. 120v.
39 Garcilaso de la Vega, *La Florida del Ynca*, f. 68r.
40 Garcilaso de la Vega, *The Florida of the Inca*, 233; Garcilaso de la Vega, *La Florida del Ynca*, f. 124r: "*reboluio a recebirle con vna flecha puesta en el arco*".
41 Garcilaso de la Vega, *The Florida of the Inca*, 234; Garcilaso de la Vega, *La Florida del Ynca*, f. 124r: "*admirados que vn animal tan animoso, feroz y brauo qual es vn cauallo, vuiesse muerto tan repentinamente de la herida de sola vna flecha, tirada de tan cerca*".
42 Garcilaso de la Vega, *The Florida of the Inca*, 234; Garcilaso de la Vega, *La Florida del Ynca*, f. 124r. "*abrieron el cauallo, y hallaron que la flecha auía entrado por los pechos y passado por medio del coraçon y buche, y tripas y parado en lo vltimo de los intestinos*".
43 Garcilaso de la Vega, *The Florida of the Inca*, 234; Garcilaso de la Vega, *La Florida del Ynca*, f. 124v: "*piden arcos y flechas, y quando no se las dan, ellos mesmos las hacen*".
44 Garcilaso de la Vega, *The Florida of the Inca*, 234; Garcilaso de la Vega, *La Florida del Ynca*, f. 124v: "*tres y quatro y seys oras con su flecha puesta en el arco aguardando con la mayor atención que se puede imaginar*".
45 Garcilaso de la Vega, *The Florida of the Inca*, 235; Garcilaso de la Vega, *La Florida del Ynca*, f. 124v. "*una cuera de ante y otra de malla que lleuaua debajo*".
46 Garcilaso de la Vega, *The Florida of the Inca*, 235; Garcilaso de la Vega, *La Florida del Ynca*, f. 125r: "*paso la cota y el cesto tan de claro, y con tanta furia, que si de la otra parte topara vn hombre tambien lo passara*".
47 Garcilaso de la Vega, *The Florida of the Inca*, 235; Garcilaso de la Vega, *La Florida del Ynca*, f. 125r: "*auer si era hombre para pasarlas ambas*".
48 Garcilaso de la Vega, *The Florida of the Inca*, 235–36; Garcilaso de la Vega, *La Florida del Ynca*, f. 125r: "*por medio del çesto, y passo los quatro doblezes que tenia la malla, y quedo la flecha atrauesada tanto de vn cabo como de otro*".
49 Garcilaso de la Vega, *The Florida of the Inca*, 236; Garcilaso de la Vega, *La Florida del Ynca*, f. 125v: "*olandas de Flandes*".
50 Núñez Cabeza de Vaca, "La Relación", 66–67: "*Dexo aquí de contar esto más largo porque cada uno puede pensar lo que se passaría en tierra tan estraña*".
51 Garcilaso de la Vega, *The Florida of the Inca*, 159, Garcilaso de la Vega, *La Florida del Ynca*, f. 83r: "*me hallo con falta de palabras necessarias para contar y poner en su punto las verdades, que en la historia se me ofrecen*".

Bibliography

Primary sources

Garcilaso De la Vega, Inca. *La traduzion del indio de los tres Dialogos de Amor de Leon Hebreo…* Madrid: En casa de Pedro Madrigal, 1590.

Garcilaso De la Vega, Inca. *La Florida del Ynca. Historia del adelantado Hernando de Soto, Gouernador y capitan general del Reyno de la Florida, y de otros heroicos caualleros Españoles è Indios; escrita por el Ynca Garcilasso de la Vega, capitán de su Magestad, natural de la gran ciudad del Cozco, cabeça de los Reynos y prouincias del Peru.* Lisboa: Impresso por Pedro Crasbeeck, 1605.

Garcilaso De la Vega, Inca. *The Florida of the Inca.* Translated by John Varner and Jeannette Varner. Austin: Texas University Press, 1951.

Núñez Cabeza De Vaca, Álvar. *La Relacion y Comentarios del Gouernador Alvar Nuñez Cabeça de Vaca, de lo Acaescido en las dos Jornadas que Hizo a las Indias.* Valladolid: Francisco Fernandez de Cordoua, 1555.

Núñez Cabeza De Vaca, Álvar. "La Relación Que dio Álvar Núñez Cabeza de Vaca de lo Acaecido en las Indias en la Armada Donde iba por Gobernador Pánfilo de Narváez, Desde el año Veinte y Siete Hasta el año de Treinta y Seis que Volvió a Sevilla con Tres de su Compañía". In *Alvar Núñez Cabeza de Vaca. His Account, His Life, and the Expedition of Pánfilo de Narváez. Volume 1*, edited by Rolena Adorno and Patrick Charles Pautz. Lincoln and London: University of Nebraska Press, 1999: 14–279.

Literature

Adorno, Rolena. "History, Law, and the Eyewitness: Protocols of Authority in Bernal Díaz del Castillo's *Historia Verdadera de la Conquista de la Nueva España*". In *The Project of Prose in Early Modern Europe and the New World*, edited by Elizabeth Fowler and Roland Greene, 154–75. Cambridge: Cambridge University Press, 1997.

Adorno, Rolena. "De Guancane a Macondo: *La Florida del Inca* y los Albores de la Literatura Latinoamericana". In *Franqueando Fronteras: Garcilaso de La Vega y La Florida del Inca*, edited by Raquel Chang-Rodríguez, 149–79. Lima: Fondo Editorial de la Pontificia Universidad Católica del Perú, 2006.

Bartsch, Shadi, and Jaś Elsner. "Introduction: Eight Ways of Looking at an Ekphrasis". *Classical Philology* 102, no. 1 (2007): i–vi.

Bauer, Ralph. *The Cultural Geography of Colonial American Literatures: Empire, Travel, Modernity*. Cambridge: Cambridge University Press, 2003.

Becker, Andrew Sprague. *The Shield of Achilles and the Poetics of Ekphrasis*. Lanham: Rowman & Littlefield Publishers, 1995.

Carey, Daniel. "The Problem of Credibility in Early Modern Travel". *Renaissance Studies* 33, no. 4 (2019): 524–47.

Chartier, Roger. "L'Imprimé et ses Pouvoirs (XVe-XVIIIe Siècles)". In *L'Imprimé et ses Pouvoirs Dans les Langues Romanes*, edited by Ricardo Saez, 21–37. Rennes: Presses universitaires de Rennes, 2010.

Cummins, Tom. "From Lies to Truth: Colonial Ekphrasis and the Art of Crosscultural Translation". In *Reframing the Renaissance. Visual Culture in Europe and Latin America, 1430-1630*, edited by Claire Farago, 152–74. New Haven and London: Yale University Press, 1995.

D'Angelo, Frank J. "The Rhetoric of Ekphrasis". *Journal of Advanced Composition* 18, no. 3 (1998): 439–47.

Daston, Lorraine, and Katharine Park. *Wonders and the order of nature, c. 1150-1750*. New York: Zone Books, 1998.
Elsner, Jaś. "Art History as Ekphrasis". *Art History. Journal of the Association of Art Historians* 33, no. 1 (2010): 10–27.
Francis, James A. "Metal Maidens, Achilles' Shield, and Pandora: The Beginnings of 'Ekphrasis'". *American Journal of Philology* 130, no. 1 (2009): 1–23.
Ginzburg, Carlo. *Threads and Traces: True, False, Fictive*. Berkeley: University of California Press, 2012.
González Echevarría, Roberto. *Myth and Archive. A Theory of Latin American Narrative*. Cambridge and New York: Cambridge University Press, 1990.
Grafton, Anthony. *New Worlds, Ancient Texts: The Power of Tradition and the Shock of Discovery*. Cambridge: The Belknap Press of Harvard University Press, 1992.
Greenblatt, Stephen. *Marvelous Possessions. The Wonder of the New World*. Chicago: University of Chicago Press, 1991.
Hartog, François. *The Mirror of Herodotus: The Representation of the Other in the Writing of History*. Translated by Janet Lloyd. Berkeley and Los Angeles: University of California Press, 1988.
Kagan, Richard L. *Clio and the Crown. The Politics of History in Medieval and Early Modern Spain*. Baltimore: Johns Hopkins University Press, 2009.
Kwiatkowski, Nicolás. "Palabra e imagen en la modernidad temprana. Representaciones de la barbarie mediante ilustración, écfrasis y alegoría". *Boletín de Estética* XII, no. 39 (2017): 43–94.
Miró Quesada, *Aurelio*. El Inca Garcilaso. Lima: Pontificia Universidad Católica del Perú, 1994.
Mora Valcárcel, Carmen de. "En torno a las ediciones de La Florida del Inca". In *Franqueando fronteras: Garcilaso de La Vega y La Florida del Inca*, edited by Raquel Chang-Rodríguez, 214–33. Lima: Fondo Editorial de la Pontificia Universidad Católica del Perú, 2006.
Myers, Kathleen A. "History, Truth and Dialogue: Fernández de Oviedo's Historia General y Natural de Las Indias (Book XXXIII, Chapter LIV)". *Hispania* 73, no. 3 (1990): 616–25.
Pagden, Anthony. *European Encounters with the New World: From Renaissance to Romanticism*. New Haven and London: Yale University Press, 1993.
Pimentel, Juan. *Testigos del Mundo. Ciencia, Literatura y Viajes en la Ilustración*. Madrid: Marcial Pons, 2003.
Preminger, Alex, and Terry V. F. Brogan, eds. *The New Princeton Encyclopedia of Poetry and Poetics*. Princeton: Princeton University Press, 1993.
Shapin, Stephen. *A Social History of Truth. Civility and Science in Seventeenth-Century England*. Chicago: University of Chicago Press, 1994.
Slater, John. "History as an Ekphrastic Genre in Early Modern Spain". *MLN* 122, no. 2 (2007): 217–32.
Van Groesen Michiel and Johannes Müller. "Introduction. Distance, Credibility, and European Geographies of Information, 1450–1750". In *Far From the Truth. Distance, Information, and Credibility in the Early Modern World*, 1–18. London: Routledge, 2023.
Voigt, Lisa. *Writing Captivity in the Early Modern Atlantic: Circulations of Knowledge and Authority in the Iberian and English Imperial Worlds*. Chapel Hill: University of North Carolina Press, 2009.

3

A COUNTRY FOR OLD MEN?

Wondrous Gerontocracy in Sir Walter Ralegh's *The Discoverie of Guiana*

William Glover

> O *Incredulitie*, the wit of Fooles,
> That slouenlie will spit on all thinges faire,
> The *Cowards castle* and the *Sluggards cradle*
> How easie t'is to be an Infidell?
>
> *George Chapman, De Guiana, carmen Epicum (1596)*[1]

On 6 February 1595 (a Thursday), Sir Walter Ralegh set sail from England with four ships, charting a south-westerly course. After several weeks at sea, the expedition reached Trinidad, at the time a Spanish colony. A skirmish with the local garrison led to the successful capture of the governor, Antonio de Berrío, who gave Ralegh valuable information about the interior geography of the mainland (present-day Venezuela). Following de Berrío's advice, along with a host of rumors and legends, Ralegh and a small crew of adventurers embarked up the lower Orinoco in small boats, in search of "the great and Golden Citie of Manoa (which the Spanyards call El Dorado)".[2]

In all this, Ralegh had both gold and royal favor on his mind. After a decade as Elizabeth I's favorite courtier—a period marked by the accrual of titles and lands, the sponsorship of journeys of exploration to the New World, and the penning of some remarkable poetry—Ralegh's star had waned.[3] A hasty, unauthorized marriage in 1591 with Elizabeth Throckmorton, one of the Queen's ladies-in-waiting, led to his banishment from the court. After lying low for a few years, Ralegh hoped to win back, at least partially, Elizabeth's esteem—"the least tast of the greatest plentie formerly possessed"—with a stunning feat of derring-do.[4] Successfully finding El Dorado would, he hoped, give the

DOI: 10.4324/9781003506928-3

English (late-comers to New World colonialism) access to vast mineral wealth, while at the same time depriving Spain of an additional source of the "Indian Golde that indaungereth and disturbeth all the nations of Europe".[5] As Benjamin Schmidt, a modern editor of Ralegh, puts it, "Guiana occupied a prime spot in the European imagination as a locus of boundless riches and imperial opportunity; it induced almost mystical (if not mythical) optimism".[6] For Ralegh, in particular, Guiana was "a locus of redemption".[7]

In the end, the expedition was (literary byproducts aside) a failure. Although Ralegh claimed to have made a number of important alliances with local Indigenous chiefs and to have reached the frontier of El Dorado, he was forced to turn back due to low supplies, poor morale, and bad weather, and emerged from the jungle bearing little but tall tales and "gold-colored stones of questionable value".[8] The voyage, then, yielded no meaningful treasure—and even Ralegh's reputation was not fully restored by it. In a prefatory epistle addressed to Charles Howard and Sir Robert Cecil, Ralegh attempts to refute rumors that he had not even been to Guiana, but had instead spent the time "hidden in Cornewall".[9] And, despite direct entreaties to Elizabeth, a full-scale invasion of the region was clearly considered unfeasible—any attempt at conquering Guiana would have to be funded and pursued by "private venturers without state support".[10] Perhaps because of these underwhelming concrete results, Ralegh's written account of his journey, *The Discoverie of the Large, Rich, and Bewtiful Empyre of Guiana* (published in 1596, shortly after his return to England), is all the more significant. Printed after Elizabeth had refused to back the sending of "but a smal army" to the region, the published text represents a pivot toward other potential investors and actors in Ralegh's Guiana project—all the while attempting to leave the door open in case of a change of heart on the Queen's part.[11] In this, Ralegh had to walk a fine line; he needed to tantalize readers with the potential for wealth and glory, and at the same time present the endeavor as feasible, and himself as a sober, reliable narrator. Guiana needed to be both "wondrous" *and* real enough to make the immense effort and risk of finding it worthwhile.

The idea of "wonder" is, of course, central to this volume, which examines it across a large time frame. Doubtless, the meaning will shift depending on context, and each author will have their own working definition. For my own chapter, it is important to define wonder's relationship with fact and fiction. Whether the term refers to an object—which could be natural (Niagara Falls) or man-made (the Pyramids)—or to the emotion elicited when encountering something "wondrous", the fundamental *perceived reality* of the experience is important. In this chapter, wonder is not a miracle, in which the laws of nature are flagrantly broken; rather, wonder stems from the "experience of the novel or unexpected" combined with "ignorance of the cause".[12] Wondrous objects, persons, or species—"the presence of radical difference", as Stephen Greenblatt puts it—certainly stretch the boundaries of the possible, but their reality is either believed or at least temporarily accepted (even if it is subsequently disproved).[13]

When it came to crafting a narrative that provoked wonder within contemporarily credible limits, Ralegh was able to draw on a long and rich tradition of European writing, in which "the margins of the world", i.e. Asia and Sub-Saharan Africa, were seen as fertile ground for "novelty, variety, and exuberant natural transgression".[14] Of course, the world's "margins" expanded greatly after 1492; the Americas "enriched [Europe's] vocabulary of wonders", and many pre-existing ones were re-contextualized in the New World.[15] Ralegh, for example, makes frequent mention of "Amazones"—female warriors from classical myth—and of a race of people with "eyes in their shoulders, and their mouths in the middle of their breasts".[16] Their name, "Ewaipanoma", may be Indigenous to the region, but they are clearly a recycling of classical *Acephali* (headless men—also known as "Blemmyes") found in, among other sources, Herodotus, Mandeville, and the margins of medieval maps.[17]

While these more familiar wonders are discussed in more detail below, the focus of this chapter is a less obvious one: old age. Put simply, Ralegh's *Discoverie* narrative contains a surprising number of aged Indigenous figures (some of them almost impossibly so), described in great detail and almost always in glowingly positive terms. Despite this, there is almost no scholarly work addressing Ralegh's gerontic emphasis; this feature of Ralegh's text is either ignored or briefly alluded to in passing, without sustained analysis.[18] In what follows, I lay out the extent of Ralegh's emphasis on extreme old age and offer some potential explanations for it. Aged characters are important as informants and guides in Ralegh's narrative, and they may also be designed to add some exoticism and color to the account. As Neal Whitehead, one of the few to briefly address the topic, suggests:

> we might consider that these reported numerical ages refer less to individual chronology than to represent... native leaders as sagacious and credible. It could also be that the discovery of *Macrobi* (long-lived or gigantic beings) aids in the textual task of establishing a convincing alterity, as was the case for other of the European marvels of encounter with America.[19]

Ultimately, however, I argue that Ralegh's account of a wondrous gerontocracy in Guiana is a finely calibrated strategic move, one that seeks both to make the conquest of El Dorado seem more feasible and to re-direct gerontophobic frustrations that were developing in the Elizabethan court.

With age comes wisdom

First and foremost, Ralegh values aged Indigenous characters for their knowledge. Like all explorers, he relies on local guides. He is candid about needing their help to navigate the labyrinthine waterways of Guiana and to explain the complex geo-politics of the region. In this, his preference is emphatically for the elderly. As he puts it,

> I sought out al the aged men, & such as were greatest trauelers, and by the one & the other I came to vnderstand the situations, the riuers, the kingdoms from the east sea to the borders of Peru, & from Orenoque southward as far as Amazones or Maragnon.[20]

Unfortunately, such aged and experienced guides are not always available. Ralegh's first guide (Ferdinando) becomes quickly lost due to a lack of knowledge, something Ralegh blames on his youth (or rather his youth when he first made the voyage in question): "of that which we entred he was vtterly ignorant, for he had not seene it in twelue yeeres before, at which time he was very young, and *of no iudgement*".[21] This guide's poor judgment manifests itself again when he is almost captured and executed by another tribe for bringing "a strange nation into their territorie"—i.e. Ralegh and his men.[22] He narrowly escapes capture, and in the ruckus, Ralegh takes hostage "a very old man", with a view to exchanging him for Ferdinando. In fact, the elderly hostage proves invaluable thanks to his knowledge of the area—"better than any stranger"—and is credited by Ralegh as saving him and his party: "but for this chance I thinke we had neuer founde the way either to Guiana, or backe to our ships".[23] The theme recurs later in the journey, when the "Lord" of the town of Arowacai lends Ralegh the services of "an old man for Pilot, who was of great experience and trauell, and knew the riuer most perfectly both by day and night".[24] This guide also provides Ralegh with ethnographic and military knowledge, such as the location of the "Aroras", the "most dangerous of all nations", on account of their poison arrows.[25] In case the reader were still in any doubt as to the value of the memory, knowledge, and judgment of the elderly Guianans, Ralegh presents their aid as a *necessity* for any future voyage: "it shall be requisite for any man that passeth" the Orinoco River "to haue such a Pilot".[26]

It may be objected that the above is hardly wondrous—the wisdom of the elderly is not a novel trope. Yet taking the emphasis placed on aged guides seriously alerts us to other, more noteworthy encounters. In Arowacai, shortly before taking on his second aged pilot, Ralegh describes a group of elders: in "that place I sawe very aged people, that we might perceiue all their sinews and veines without any flesh, and but euen as a case couered onely with skin".[27] The detail is striking, and the abrupt sequence of "and but euen", which pauses the sentence's flow as Ralegh hunts around for an appropriate simile, speaks to his breathless fascination with the sight of these fragile, aging bodies.

This fascination will be further stimulated by two of the most important characters in the narrative: Carapana and Topiawari. Both are Indigenous leaders—*caciques*, to use the Hispanicized Taino term. Both are exceedingly shrewd, physically strong, and a hundred years old (or more). Ralegh is aware of their advanced age and bodily vigor even before he meets them, having heard of their mistreatment by the Spanish, who allegedly "made the ancient Casiqui which were Lordes of the country to be their slaues", and "kept them in chains, & dropped their naked bodies with burning bacon, & such other torments".[28] In emphasizing the physical torment and indignity suffered by these figures, Ralegh crafts compelling anti-Spanish rhetoric: cruel disrespect of venerable elders is added to an already lengthy rap sheet dominated by violence, rapine, and Catholicism. By comparison, Ralegh is able to present himself as gracious and respectful, bestowing praise on those mistreated by his enemies:

> The king of this land is called Carapana, a man very wise, subtill, and of great experience, being little lesse than 100. yeeres old... Carapana kept himselfe and his country in quiet and plenty: he also held peace with Caribas or Canibals his neighbors, and had free trade with all nations whosoeuer els had war.[29]

Carapana's canniness is also a source of frustration for Ralegh, who refers to him as an "olde fox". The *cacique* refuses to meet with the English, doubtless in order to maintain the aforementioned neutrality, though this does not stop Ralegh from assuring his readers (not entirely convincingly) of their closeness:

> in mine opinion, or rather I assure my selfe, that Carapana (beeing a notable wise and subtile fellowe, a man of one hundred yeares of age, and therefore of greate experience) is remooued, to looke on [i.e. standing back to observe], and if hee finde that wee returne strong, hee will be ours.[30]

Ralegh has far closer relations with Topiawari (another *cacique*, and the uncle of Morequito, a leader deposed by the Spanish), with whom he meets on multiple occasions. More than any other character in the *Discoverie* narrative, Topiawari's age and physical strength stretch the bounds of credibility. He survives a seventeen-day forced march at the hands of the Spanish before being ransomed and released, and when he first meets with Ralegh, he travels "on foote from his house, which was 14 English miles", and "returned on foote the same daie"—all this at the ripe age of "110. yeers old".[31]

It should now be clear that aged figures crop up repeatedly in this short narrative and are of great importance when they do. They may range on the "wonder" spectrum—some, like Topiawari, are more surprising than others—but there is something undeniably striking about the sheer frequency with which they appear. This all begs a simple question: why? What purpose do these figures serve in the *Discoverie*? No travel writer is a neutral reporter—especially not a would-be colonist like Ralegh. Whether he really encountered these figures is almost beside the point. Rather, the critic must focus on why Ralegh chose to foreground them in his text as much as he did—and why he focused so heavily on the numerical ages of Carapana and Topiawari. To be sure, we should acknowledge that Ralegh and his Indigenous interlocutors may well not have shared a numerical or calendric system that would allow personal information like age to be adequately translated. And, even if the two parties did see eye to eye on age and how it was measured, it may be that Carapana and Topiawari simply lied to Ralegh, perhaps in an attempt to overawe him with their longevity and strength. It is doubtful, though, that such a canny operator as Ralegh would have been taken in by this. His nickname for Carapana—"olde fox"—suggests that he is aware of the possibility of deception. And, of course, as author, Ralegh was firmly in control. He may have been given the numbers; he may have made them up. But regardless, he must have had his own reasons for including them in the account and for emphasizing them as much as he does. He clearly wanted his readers to believe that he genuinely had encountered these figures, and that they were genuinely as old as he claimed them to be. Even if they were invented wholesale (though most editors seem confident that they were not), there must have been, in Ralegh's words, "profit in the imagination".[32]

As Whitehead argued earlier, Carapana and Topiwari's advanced age certainly bolsters the reader's sense of their knowledge and wisdom. We could also read accounts of ancient, vital figures as one classic "wonder" among many others—*Macrobi*, to go along with Amazons and *Acephali*. Yet Ralegh is notably cautious when discussing these more traditional wonders. Rather than claiming first-hand knowledge, he takes care to always place the onus of his truth-claims on a carefully cited secondary source. When discussing the Amazons, for example, he acknowledges that "the truth of those warlike women" is disputed: "of some it is beleeued, of others not".[33] He goes on to describe their customs in detail, drawing on the testimony of a friendly "Casique or Lord of people" rather than first-hand experience.[34] The resulting description balances open-mindedness with skepticism, reinforcing certain conventions while dispelling others: the Amazons may indeed be "cruell and bloodthirsty" warriors, and if "deliuered of a sonne, they returne him to the father"—yet that they "cut of the right dug of the breast" Ralegh does "not finde to be true".[35] The Ewaipanoma (Blemmye-like headless men) are similarly handled. Their existence is attested to by the son of a friendly chief, along with "euery child in the prouinces of Arromaia and Canuri", though perhaps this was felt to be insufficient, as Ralegh added a reference to John Mandeville to the published text:

> [s]uch a nation was written of by Maundeuile, whose reports were held for fables many years, and yet since the East Indies were discouered, we finde his relations true of such things as heereteofore were held incredible.[36]

Ralegh is "resolued that so many people did not all combine, or forethinke to make the report"—but if they did, his framing of the evidence makes him, at worst, gullible rather than an outright liar.[37]

The same dynamic can be seen even in the primary purpose of Ralegh's journey. The existence of Manoa is confirmed by a deceased Spaniard, "Iohannes Martines", whose written "relation" can apparently be read "in the Chauncery of Saint Iuan de Puerto rico"—but isn't, at least not by Ralegh.[38] The reality of the golden city is never questioned, and the closest the English get to it is a town allegedly "adioyned to" a "frontier towne of the Empire".[39] The question of gold itself is also carefully negotiated: it is supposedly abundant, attested to by Spanish and Indigenous alike, but is always just out of reach due to distance or a lack of proper supplies. For example, Ralegh complains that he "had neither pioners, bars, sledges, nor wedges of Iron", yet simultaneously insists that "euery stone that we stooped to take up, promised eyther golde or siluer by his complexion".[40]

In each of these cases, then, Ralegh retains a degree of plausible deniability—as to be expected in a genre where the narrator's credibility is both essential and, by default, suspect.[41] When it comes to his encounters with the aged, however, Ralegh claims first-hand knowledge of their existence and their vitality, and stands by the numerical ages he assigns to them. This brings us back to my previous emphasis on the perceived reality of wondrous objects or people. Ralegh's "chiefest friend", the 110-year-old Topiawari, is a wondrous figure, and the repeated invocation of his age is clearly designed to shock; but at the same time, Ralegh clearly wants his readers to take his existence *literally*, and not as a traveller's tale.[42] It should here be noted that a healthy 110-year-old would have been seen by Ralegh's audience as incredible and noteworthy, but not strictly impossible. Keith Thomas has pointed out "the remarkable number of supposed centenarians" in early modern England, "ages [that] were accepted uncritically by most contemporaries".[43] One Thomas Parr, for example, supposedly lived to be 152 and was dissected by the royal physician William Harvey upon his death. Harvey accepted Parr's age, ascribing his longevity not to anything miraculous, but rather the "perfect purity" of rural air, and a lifetime of "very simple" food.[44]

If the mere existence of these figures is not itself quite as noteworthy as a modern reader might presume, perhaps their significance lies in the elevated social position they occupy. Travelogues have long been employed to criticize failings in the author's home nation, and there is literary precedence for this exact issue of aged authority.[45] Thomas More located his *Utopia* (1516) off the coast of Brazil—not too far from Ralegh's Guiana—and, in imagining a well-ordered state, saw the elderly as playing an essential role. Each household is run on a gerontocratic basis: "[t]he eldest… ruleth the family. The wives be ministers to their husbands, the children to their parents, and, to be short, the younger to their elders".[46] In Utopia, whole communities dine together, and presiding over each meal, alongside the head of the ward and his wife, are "two of the ancientest and eldest".[47] Among the regular townsfolk, the seating arrangements and order of service also reflect the primacy of the old:

> On both sides of them sit young men, and next unto them again old men… This, they say, was ordained to the intent that the sage gravity and reverence of the elders should keep the younger from wanton license of words and behaviour… The dishes be not set down in order from the first place, but all the old men… be first served of their meat… Thus the elders be not defrauded of their due honour.[48]

Care for the elderly is a moral imperative: "they judge it a great point of cruelty that anybody in their most need of help and comfort should be cast off and forsaken, and that old age, which both bringeth sickness with it and is a sickness itself, should unkindly and unfaithfully be dealt withal".[49] By contrast, as the narrator Hythloday points out, the outlook of an aging laborer in early modern Europe was a desperate one: "the remembrance of their poor, indigent, and beggarly old age killeth them up. For their daily wages is so little that it will not suffice for the same day, much less it yieldeth any overplus that may daily be laid up for the relief of old age".[50]

But sometimes age comes alone

More's emphasis on the precarity of old age gets us closer to what I argue to be the root of Ralegh's aged portraits. As we have established, the most noteworthy Indigenous leaders in Guiana are extremely old—and, as Ralegh insists throughout, friendly to the English. Yet, because of their age and the threatening presence of the Spanish, their authority is unsteady. Topiawari, again, provides an important case study; despite his physical vitality, he cuts a deeply troubled figure. As his mistreatment by the Spanish demonstrates, his authority is far from assured, and his succession is threatened. In recounting the state of his kingdom, "he answered with a great sigh (as a man which had inward feeling of the losse of his conntrey and liberty, especially for that his eldest sonne was slain in a battel…, whom he most entirely loued)".[51] This loss of youth and potential in the royal family seems to be mirrored among the general populace of the "borderers"—the name Ralegh gives to the Indigenous nations that supposedly surround El Dorado, and that he hopes to ally himself with: "whereas they were wont to haue ten or tweluc wiues, they were now inforced to content themselues with three or fower".[52] As a result of this, as well as wartime depopulation, "those frontiers are growen thinne of people, and also great numbers are fled to other nations farther off for feare of the Spanyardes".[53] The health of monarch and people are intertwined, and Topiawari, it seems, is not long for this world, as one goodbye to Ralegh indicates: "he desired leaue to depart, saying that he had far to go, that he was old, & weake, and was euery day called for by death, which was also his owne phrase".[54] This grim picture makes Topiawari's decision to put his only remaining son into the care of Ralegh even more striking and shows the desperation of his position: "he freely gaue me his onelie sonne to take with me into England, and hoped, that though he himselfe had but a short tyme to liue, yet that by our meanes his sonne should be established after his death".[55]

The venerable Topiawari, then, is also vulnerable—and it is here that the specific tenor of Ralegh's own public image becomes crucial. Imperialists tend to cloak themselves in euphemistic rhetoric, and Ralegh was no exception. He was a tireless proponent of what would become known as "The Black Legend"—the portrayal of Spain as "intolerant, brutal and fanatical"—and sought to frame his own conduct as a positive contrast.[56] Throughout the *Discoverie*, Ralegh presents himself as a faithful friend to the Indigenous nations he encounters, a powerful ally against Spain, and generally an upstanding figure. Forestalling accusations of impropriety, for example, he avers that neither he nor his men "euer knew any of their women, and yet we saw many hundreds, and had many in our power, and of those very young, and excellently fauored which came among vs without deceit, starke naked".[57] Historians are divided on how seriously we should take Ralegh's moral posturing. Anthony Esler sees little difference between Ralegh and his enemies, arguing that the Englishman "had developed a great admiration for and desire to emulate the Spanish conquistadors".[58] This goes a little far, ignoring as it does Ralegh's vociferous denunciations of Spanish brutality. In the opposite camp is one of Ralegh's recent biographers, Alan Gallay, who suggests that Ralegh "envisioned an empire without conquest, where the Native peoples would be full partners in the colonial enterprise"—in short, "Ralegh proposed… a benevolent empire".[59]

While this may have been Ralegh's public stance, he was far more direct in private, as evidenced in a government document pertaining to Ralegh's journey, "Of the Voyage for Guiana".[60] Some scholars ascribe it directly to Ralegh himself, others to his associates, Lawrence Keymis and Thomas Hariot.[61] Either way, the "treatise [is] clearly intended to hammer home the feasibility, practicality and overriding importance of Ralegh's New World project", and lays out a *realpolitik* rationale for a full-scale military expedition to Guiana.[62] While it criticizes the Spanish for their "barbarous & exquisite Massacres", it also argues in no uncertain terms that "subduing" and "annex[ing]" Guiana and its surrounding kingdoms "to the Crowne imperiall of the Realme of England" would be "honorable, profitable, necessary" and accomplished "with no greate chardge, or difficultye".[63] Ralegh may well offer friendship to *caciques* like Carapana and Topiawari, but his ultimate aim is, if not to depose them, then to reduce them to a state of vassalage. This is confirmed by Kemys, who in his *Relation* bluntly reminds Ralegh of his reasoning for befriending

Topiawari: "you had left this aged Sire aliue, to bee a blocke in their way, to whome after his decease, this enterprise by patent did belong, and to be a weake aduersarie against your self, whom at all times, you knew easilie how to distresse".[64] The friendly, aging *cacique*, whose son and successor (it should be remembered) is in Ralegh's power, should thus be read not as merely a curiosity, but as a tempting soft target for an expansionist England.

Most importantly, we must bear in mind the attitudes toward age and aging in Elizabeth's court in the mid-1590s. The Queen's long reign, close to the end of its fourth decade by the time of Ralegh's Guianan voyage, saw rising tensions between old and young; enduring gerontocratic rule clashed with the frustrations of a younger generation of upwardly mobile courtiers and military leaders.[65] Anthony Esler has written at length on the frustrations of the "Elizabethan younger generation".[66] More recently, Christopher Martin has described a "fierce generational strife", one that culminated with the execution of the rebellious Earl of Essex in 1601.[67] This event

> marked the culmination—for better or for worse—of a grand generational conflict that had taken shape over the 1590s… the thirty-three-year-old earl embodied the restlessness of a social stratum impatient for promotion before the attenuated but lingering gerontocracy concentrated in the person of the aging queen herself.[68]

Although, in 1595, this bloody climax was still some way off, tensions would have been evident. Esler portrays Ralegh as a part of this bullish "Marlovian" generation, but in fact, he was somewhere in the middle of the brewing conflict: around twenty years younger than Elizabeth, but about ten years older than Essex.[69] Thus, it may be that Ralegh also felt the younger courtiers pressing at his back, particularly during his exile from court. This goes some way to explaining some of the last, and most important, lines in the *Discoverie*. In his final, exhortatory address to the Queen herself, after assuring her of the weakness of the Spanish and the "loue" of the natives, Ralegh advises: "Her maiestie may in this enterprize employ *all those souldiers and gentlemen that are yonger brethren*, and all captaines and Chieftaines that want employment".[70] Here, Ralegh concisely posits the conquest of Guiana as a solution to the problem of an overly ambitious, under-employed youth, both in the army and the court.[71] In a letter to Charles Howard, Lord High Admiral, he makes an even more overt appeal to worries the Queen might have about her restless, youthful courtiers:

> Her Majestye shall by foreslowing [the Guiana enterprise] lose the greatest assurance of good that ever was offred to any Christien prince & your Lordshipp douth fynde that it is the surest way to devert all attempts from home.[72]

The key word here is "attempts", almost certainly used in the now obsolete sense of "a warlike enterprise; an attack, assault, onset".[73] The invasion of Guiana, and the displacement of its elderly chieftains (who, isolated and devoid of heirs, are uncomfortably reminiscent of Elizabeth herself) could serve as a convenient opportunity to re-direct—or sublimate—the frustrated energies of a young, hot-headed generation, and help to defend the realm against its adversaries, both foreign and domestic.[74]

There is one final context to bear in mind: the practice of contrasting the Americas and Europe with a vocabulary of age: the "New" World vs. the "Old". This was not incidental or neutral terminology—rather, it was part of European colonialism's discursive foundation. In a famous passage toward the end of his account, Ralegh's summarizes his impressions of Guiana with a metaphor of sexual immaturity: "[t]o conclude, Guiana is a Countrey that hath yet her Maydenhead, neuer sackt, turned, nor wrought… the graues haue not been opened for gold, the mines not broken with sledges".[75] The landscape of this youthful "New World", with its unspoiled beauty and untapped mineral wealth, contrasts sharply with an over-crowded, violent Europe racked by wars and religious conflict.[76] At times, this youthfulness is also ascribed by colonial writers to Indigenous people, in the sense of being child-like or naïve. Thomas Hariot, an associate of Ralegh's, presents Native people as dumbstruck in the face of European technologies, such as "sea compasses… gunnes… writing and reading", which "so far exceeded their capacities to comprehend" that "they thought they were rather the works of gods than of men".[77] Later, he explicitly invokes the perceived age of European settlers: "[s]ome were… of opinion that we were not born of women, and therefore not mortal, but that we were men of an old generation many years past, then risen again to immortality".[78] In the case of Guiana, however, the script is flipped: the Indigenous leaders there are canny and old—wondrously so, in some cases—but consequently weak and unsteady, and ripe for displacement by the more vigorous and youthful English.[79]

Out with the old, in with the new

With these various contexts in mind, the preponderance of elderly figures in the *Discoverie* narrative, and the special significance of Carapana and Topiawari, begin to make sense, allowing Ralegh, as they do, to perform a complex series of rhetorical maneuvers. While the *caciques*' wisdom and knowledge help to bolster the legitimacy of Ralegh's account and make them useful allies in the short term, their advanced age and the fragility of their kingdoms make their long-term prospects of holding onto power doubtful. The Edenic, virginal land of Guiana is presided over by the withered hand of old age—as Topiawari says, he is "euery day called for by death".[80] He and his peers can therefore easily (and peaceably) be pushed aside when the time is right, allowing the English to accrue colonial power and mineral wealth, and still maintain claims of moral superiority over the Spanish.[81] Finally, this imminent power vacuum in Guiana provides an outlet for England's "second brethren": ambitious, under-employed, and chafing against aged rule in their own country. The myth of El Dorado captured Europe's imagination because of its wealth, and in his attempt to garner support for a military expedition to conquer it, Ralegh dwells at length on gold, silver, and precious stones. At the same time, however, he mobilizes a less obvious wonder—old age—which proves a potent rhetorical device, at once bolstering Guiana's wondrous reputation, and the hope that the English might one day conquer it.

Notes

1 Chapman, "De Guiana, carmen Epicum", in Kemys, *A Relation of the Second Voyage*, 7. A note on pagination: this edition of Kemys does not have numbered pages; the page numbers I give indicate the "image" number on EEBO, each of which refers to a scanned double-page spread.
2 Ralegh, *Discoverie of Guiana*, 1. A note on the text: in subsequent references, Arabic numerals preceded with "MS" refer to the manuscript (Lambeth Palace MS 250, ff. 315r-337v.) printed alongside the published version of the text in the facing-page edition edited by Joyce Lorimer. I have, when quoting, retained Ralegh's original spelling, but have removed his idiosyncratic italicization for clarity, and to make my own use of emphasis clearer. Finally, a note on spelling: as per early modern conventions (or lack thereof), Ralegh spelled his name both with and without an "i". While some readers may be more familiar with "Raleigh", I have followed the majority of scholarly sources in writing it as "Ralegh".
3 For a brief biography of Ralegh, see Nicholls and Williams 2022.
4 Ralegh, *Discoverie of Guiana*, 4–5.
5 Ralegh, *Discoverie of Guiana*, 14. The name "El Dorado" has a convoluted history. Initially referring to the ruler of a kingdom so wealthy that it was a custom for him to cover himself in gold dust—hence "the golden one" in Spanish—it eventually came to refer synecdochally to the city-state itself. See Lorimer 2006, xxvii-xxix for a brief summary of the Spanish expeditions to find El Dorado, and Schmidt 2008, 20–23.
6 Schmidt 2008, 21.
7 Schmidt 2008, 23.
8 Schmidt 2008, 29.
9 Ralegh, *Discoverie of Guiana*, 5.
10 Lorimer 2006, xxxviii. Admittedly, this was not out of the ordinary for the English or for Ralegh, even when his favor with the Queen was at its height. As Kathleen Donegan argues in her discussion of Elizabeth's 1584 charter to Ralegh, which grants him permission to "discouer, search, finde out, and view such remote, heathen and barbarous lands, countries, and territories" (*Charter to Sir Walter Ralegh*, 53), the Queen explicitly absolves herself of direct responsibility for the colonists and their well-being. On the contrary, "the Crown would take only the rewards" (Donegan 2014, 27) from any colonial venture. For a close analysis of the charter's language, see Donegan 2014, 25–27.
11 Ralegh, *Discoverie of Guiana*, 217. For a note on the timing of Elizabeth's rejection, see Lorimer 2006, xxxviii. For an example of Ralegh's attempts to continue to appeal to the monarch in print: when describing some of the Indigenous tribes he encounters on his journey, Ralegh (*Discoverie of Guiana*, 199) adds to the published text a significant detail, namely that the "Canuri… are gouerned by a woman (who is

inheritrix of that prouince)… and asked mee diuers questions of her Maiesty, beeing much delighted with the discourse of her Maiesties greatnes, and wondring at such reports as we truely made of her highnes many vertues".

12 Daston and Park 1998, 23. The quotation is taken from Daston and Park's analysis of the theorization of wonder by Gervase of Tilbury, a thirteenth-century English statesman and writer.

13 Greenblatt 1991, 14.

14 Daston and Park 1998, 25.

15 Daston and Park 1998, 108.

16 Ralegh, *Discoverie of Guiana*, 157. For references to "Amazones", see pages 55, 63, 65, 201, 221.

17 See Mandeville, *The Travels of Sir John Mandeville*, 137; Herodotus, *Herodotus, Books III & IV*, 2:395; "Hereford Mappa Mundi". See Lorimer 2006, lxxii-lxxiv for a brief discussion of the transferal of Classical myths to the Americas. For an account of "visual recyclings by early modern publishers" such as Theodore de Bry, see Leitch 2024, 70. For an interpretation that roots the headless Ewaipanoma in "the native politics of the region" rather than Classical myth, see Whitehead 1997b, 87–88.

18 Mary Campbell, for example, when discussing Ralegh's interaction with a 110-year-old chieftain, states that the author makes him seem "real and exotic at once, both alien and human", but does not specifically mention his age (Campbell 1988, 243). Mary Fuller acknowledges the same chieftain's age in passing but does not dwell on it (Fuller 1991, 53). Alden Vaughan's article on Ralegh's "Indian interpreters" does not mention that most of his guides are elderly (Vaughan 2002). Lengthy introductions by Joyce Lorimer and Benjamin Schmidt do not address the issue. Finally, while John Sell gives a brief interpretation of one of the aged guides, suggesting that he "doubles as a Charon figure", he moves quickly on, saying that it "need not worry us" (Sell 2006, 119).

19 Whitehead 1997a, 112.

20 Ralegh, *Discoverie of Guiana*, 73. This sentiment is also expressed on page 63: "I made inquirie amongst the most ancient and best traueled of the Orenoqueponi".

21 Ralegh, *Discoverie of Guiana*, 95 (my emphasis). Curiously, the italicized passage echoes the language of the 1573 Statute of Artificers, which expresses a similar ideology in forbidding men under the age of 24 from taking on apprentices, on the grounds that they are "withoute judgment". See Tawney and Power 1951, 1:354.

22 Ralegh, *Discoverie of Guiana*, 97.

23 Ralegh, *Discoverie of Guiana*, 97, 99. Ferdinando, the youthful but inexperienced guide, reappears in a travel account published by Lawrence Kemys—Ralegh's deputy, who returned to Guiana in 1596, the year after Ralegh's first voyage. After Kemys gives him "good hope" that the English will eventually remove the Spanish from the area, Ferdinando responds thus: "he to deserue some part of this friendship, commended vnto vs an elderlie man to bee our Pilot" (Kemys, *A Relation of the Second Voyage*, 11).

24 Ralegh, *Discoverie of Guiana*, 129.

25 Ralegh, *Discoverie of Guiana*, 135.

26 Ralegh, *Discoverie of Guiana*, 129.

27 Ralegh, *Discoverie of Guiana*, 129.

28 Ralegh, *Discoverie of Guiana*, 29.

29 Ralegh, *Discoverie of Guiana*, 75.

30 Ralegh, *Discoverie of Guiana*, 189–91. The reference to "old foxe" is on 191. Carapana's coyness recurs in Kemys's account, in which he tantalizes with the prospect of a meeting, defers it by "six or seuen daies", and eventually sends "one of his aged followers, to certifie vs, that he was sicke, olde and weake… that therefore he desired vs to hold him excused for not coming". Despite this, he re-states through

his spokesman his sympathies with the English: "[t]his olde man dilated vnto vs, that Carapana in hope of our returne, hath euer since your Lordships being in that countrie, kept the mountains, where the Spaniards can hardlie anie way inforce him" (Kemys, *A Relation of the Second Voyage*, 19).

31 Ralegh, *Discoverie of Guiana*, 171, 139. It should be noted that Topiwari's age is not strictly consistent throughout the narrative. When first mentioned in the context of his ransom, he is "aboue 100. yeeres old, and yet of a very able bodie" (83). When he later meets with Ralegh, he is "110. yeers old" (139), which itself is a subtle downgrade from the manuscript, which has him at "above 110 yeres old" (MS 138). It seems his longevity is well known to the Spanish, too—Lorimer includes in her edition a Spanish document that refers to Ralegh's relationship with "an ancient old man whom they say was close to one hundred and twenty years old, whose name was Topiwary [sic]". See "Relationship of the deceased Prince of Great Britain with Ralegh" (Lorimer 2006, Appendix V, 303).

32 Ralegh, *Discoverie of Guiana*, 157.

33 Ralegh, *Discoverie of Guiana*, 63.

34 Ralegh, *Discoverie of Guiana*, 63.

35 Ralegh, *Discoverie of Guiana*, 65, 63, 65. This last detail is absent in the MS, and so represents another strategic edit to the printed text.

36 Ralegh, *Discoverie of Guiana*, 157.

37 Ralegh, *Discoverie of Guiana*, 157. For an analysis of the rhetorical uses of the *Acephali* trope in European writing about the Americas, see Oldenburg 2008.

38 Ralegh, *Discoverie of Guiana*, 43.

39 Ralegh, *Discoverie of Guiana*, 149.

40 Ralegh, *Discoverie of Guiana*, 179, 153. For a brilliant analysis of the deferrals and gaps in Ralegh's narrative, see Fuller 1991.

41 The questions of belief and credibility are important ones for travel writing, especially in the early modern period, when it was nearly impossible for a reader to verify the claims made by a given text. For an article that examines both authorial and readerly attitudes toward belief in early modern travel writing, see Carey 2019. For a longer, more varied treatment of the subject, see Van Groesen and Müller 2024. For an account that situates Ralegh's negotiation of credibility within the history of the novel, see Johnson 2021. For an argument that Ralegh's "narrative is based on epistemological strategies that adumbrate in many ways the Baconian method" (113), see Cottegnies 2011.

42 For reference to "chiefest friend", see Ralegh, *Discoverie of Guiana*, 149.

43 Thomas 1976, 234–35.

44 Harvey, *The works*, 591–92.

45 For a book-length treatment of social critique through early modern travel writing, see Hadfield 1998.

46 More, *Utopia*, 63.

47 More, *Utopia*, 66.

48 More, *Utopia*, 66–7.

49 More, *Utopia*, 91.

50 More, *Utopia*, 120. As one scholar has pointed out, "in the sixteenth century old-age pensions were so rare that no term for it existed" (Gilbert 1967, 28).

51 Ralegh, *Discoverie of Guiana*, 143.

52 Ralegh, *Discoverie of Guiana*, 173.

53 Ralegh, *Discoverie of Guiana*, 173.

54 Ralegh, *Discoverie of Guiana*, 145.

55 Ralegh, *Discoverie of Guiana*, 177.

56 Edelmayer 2011, 2. While Edelmayer does not see the "*leyenda negra* [as] systematically vilify[ing] the Spanish" (29), he does see it as a prominent European-wide phenomenon. For an account focused on England, see Maltby 1971.

57 Ralegh, *Discoverie of Guiana*, 121.
58 Esler 1966, 177.
59 Gallay 2019, 362.
60 See Lorimer 2006, xxxv-xxxvii, and Schmidt 2008, 136–7 for brief introductions to the document. For the document itself, see "Of the Voyage for Guiana" (Appendix II, 253–63 in Lorimer 2006).
61 Schmidt suggests that the manuscript "appears to be in Ralegh's hand" (136), whereas Lorimer focuses on "stylistic resemblances to [Kemys's] *Relation*" (xxxvi), following Lefranc 1968, 58.
62 Lorimer 2006, xxxvi. This is also supported by Lefranc 1968, 58: "even if, strictly speaking, these pages are not entirely the work of Ralegh, they do reflect his thoughts" (my translation).
63 *Of the Voyage for Guiana*, 253–54.
64 Kemys, *A Relation of the Second Voyage*, 13–14.
65 Questions about ageing—the virtues and vices of the old, and their proper position in society—were debated across Europe in the early modern period. Classical works on the subject were translated and re-published, and new works were written in prose and poetry. Opinions were divided; for every writer who praised the wisdom, tact, and restraint of the elderly, another portrayed them as "peevish, forgetful, covetous, garrulous, and dirty" (Thomas 1976, 244). Old age was considered from medical, moral, philosophical, and economic angles, but it was arguably the political dimension that proved most thorny—the nub of the issue distilled by Plutarch as "whether an aged man ought to manage publike affaires". The clash between gerontocratic ideals and gerontophobic realities was particularly acute toward the end of Elizabeth's reign, when the ruler in question was not only old, but also a woman. For a small selection of primary texts that tackle the subject, see Cicero, "The Book of Old Age"; Bacon, "Of Youth and Age"; Cornaro, *Writings on the Sober Life*; Erasmus, "On the Troubles of Old Age"; Plutarch, "Whether an Aged Man Ought to Manage Publike Affaires". For an excellent scholarly monograph on the subject, see Martin 2012.
66 Esler 1966.
67 Martin 2012, 2.
68 Martin 2012, 30.
69 Esler 1966, xxiv.
70 Ralegh, *Discoverie of Guiana*, 217 (my emphasis).
71 Nor was he the only one to make this kind of argument. His contemporary, Richard Hakluyt, argued for the English colonization of America partly on the grounds that it would allow the kingdom to "unladen" "[t]he frye [children] of the wandringe beggars of England, that growe upp ydly, and [are] hurtefull and burdenous to this realme" (Hakluyt, *Discourse Concerning Western Planting*, 160–61).
72 Ralegh, *Discoverie of Guiana*, 285.
73 Oxford English Dictionary, "Attempt, n., Sense 3.a." The *OED* locates this sense precisely in the period in question, giving the years 1584–1665.
74 The issue of Elizabeth's age is directly alluded to in the George Chapman poem that begins this essay. See Riehl 2010, 159–60.
75 Ralegh, *Discoverie of Guiana*, 211. For an excellent analysis of the *Discoverie* that places this quotation at the center of its analysis of gender and sexual politics, see Montrose 1993.
76 An aura of youthful promise has since been self-consciously incorporated into the political identity of the United States. See, for example, the final lines of Emma Lazarus's 1883 poem, "The New Colossus", engraved at the feet of the Statue of Liberty, in which the youth of the New World becomes triumphalist: "'Keep, ancient lands, your storied pomp!' cries she / With silent lips. 'Give me your tired, your poor, / Your huddled masses yearning to breathe free, / The wretched refuse of

your teeming shore. / Send these, the homeless, tempest-tost to me, / I lift my lamp beside the golden door!'" (Lazarus, *The New Colossus*).
77 Hariot, "A Briefe and True Report of Virginia", 899.
78 Hariot, "A Briefe and True Report of Virginia", 902.
79 It is important to note an underlying subtext that brings together virility, age, and decorum. As Leslie A. Fielder notes, it is generally felt that the "*properly old*" ("proper" both in the sense of "truly" and "appropriately") must "withdraw from sexual competition" (Fielder 1979, 236). Guiana is, Ralegh suggests, a virginal land *indecorously* ruled over by impotent (and, therefore, feminized) elders—recall the declining population presided over by Topiawari mentioned above (p.11). The English, on the other hand, imagine themselves as sexually potent; as Chapman writes in *De Guiana, carmen Epicum*, "all our Youth take *Hymens* lightes in hand, / And fill each roofe with honor'd progenie" (Kemys, *A Relation of the Second Voyage*, 8). As for Ralegh, his repeated assertion of sexual restraint also demonstrates sexual potential, though he also attempts to garner sympathy at the outset of his narrative by describing himself as being "in the winter" of his life (Ralegh, *Discoverie of Guiana*, 5).
80 Ralegh, *Discoverie of Guiana*, 145. And indeed, Kemys records in 1595 "that Topiwarie is dead" (Kemys, *A Relation of the Second Voyage*, 17).
81 We might see in Topiawari an early version of the "vanishing Indian" trope, in which Euro-American writers portray the displacement of Native peoples as a tragic inevitability. Ralegh is not the only would-be colonist using such rhetoric at this time; his contemporary, Captain John Smith, ventriloquizes his indigenous adversary, Powhatan, as follows: "Captaine Smith, you may understand that I having seene the death of all my people thrice, and not any one living of those three generations but my selfe; I know the difference of Peace and Warre better then any in my Country. *But now I am old and ere long must die*" (Smith, "The Generall Historie of Virginia", 371, my emphasis).

Bibliography

Primary sources

Bacon, Francis. "Of Youth and Age." In *The Essayes or Counsels, Ciuill and Morall, of Francis Lo. Verulam, Discount St. Alban*, 247–50. London: John Haviland, 1625.

Charter to Sir Walter Raleigh. *In The Federal and State Constitutions, Colonial Charters, and Other Organic Laws of the States, Territories, and Colonies, Now or Heretofore Forming The United States of America*, 1:53–57. Washington, D.C.: Government Printing Office, 1909.

Cicero, Marcus Tullius. "The Book of Old Age." In *Fovvre Seuerall Treatises of M. Tullius Cicero*, translated by Thomas Newton. London: Thomas Marshe, 1577.

Cornaro, Alvise. *Writings on the Sober Life: The Art and Grace of Living Long*. Edited by Marisa Milani. Translated by Hiroko Fudemoto. Toronto: University of Toronto Press, 2014.

Erasmus, Desiderius. "On the Troubles of Old Age." In *Collected Works of Erasmus: Poems*, edited by Harry Vredeveld, translated by Clarence H. Miller, 13–25. Toronto: University of Toronto Press, 1993.

Hakluyt, Richard. *A Discourse Concerning Western Planting. Written in the Year 1584 by Richard Hakluyt*. Edited by Charles Deane and Leonard Woods. Cambridge, MA: Press of John Wilson and Son, 1877.

Hariot, Thomas. "A Briefe and True Report of Virginia." In *Captain John Smith: Writings with Other Narratives of Roanoke, Jamestown, and the First English*

Settlement of America, edited by James Horn, 874–905. New York: The Library of America, 2007.
Harvey, William. *The Works of William Harvey, M.D.* Translated by Robert Willis. London: The Sydenham Society, 1847.
"Hereford Mappa Mundi."(n.d.), Accessed June 14, 2024. https://www.themappamundi.co.uk/mappa-mundi/.
Herodotus. *Herodotus, Books III & IV*. Translated by A. D. Godley. Vol. 2. 4 vols. London: William Heinemann, 1928.
Kemys, Lawrence. *A Relation of the Second Voyage to Guiana. Perfourmed and Written in the Yeare 1596. By Lawrence Kemys, Gent*. London: Thomas Dawson, 1596.
Lazarus, Emma. *The New Colossus*, 1883.
Mandeville, John. *The Travels of Sir John Mandeville*. Edited by C. W. R. D. Moseley. London: Penguin, 2005.
More, Thomas. "Utopia." In *Three Early Modern Utopias: Utopia, New Atlantis, The Isle of Pines*, edited by Susan Bruce, translated by Ralph Robinson. Oxford: Oxford University Press, 1999.
Of the Voyage for Guiana. *BL, Sloane MS 1133*, ff.45–52. Printed in Lorimer 2006, Appendix II, 253–263.
Plutarch. "Whether an Aged Man Ought to Manage Publike Affaires." In *The Philosophie, Commonlie Called, The Morals Written by the Learned Philosopher Plutarch of Chæronea*, translated by Philemon Holland, 382–401. London: Arnold Hatfield, 1603.
Ralegh, Sir Walter. *Sir Walter Ralegh's Discoverie of Guiana*. Edited by Joyce Lorimer. London: Ashgate, 2006.
"Relationship of the deceased Prince of Great Britain with Ralegh." *From MS Biblioteca Nacional, Madrid, 1868418*. Translated and printed in Lorimer 2006, Appendix V, 303–4.
Smith, John. "The Generall Historie of Virginia, New-England, and the Summer Isles." In *Captain John Smith: Writings with Other Narratives of Roanoke, Jamestown, and the First English Settlement of America*, edited by James Horn, 199–670. New York: The Library of America, 2007.

Literature

Campbell, Mary B. *The Witness and the Other World: Exotic European Travel Writing, 400-1600*. Ithaca, NY: Cornell University Press, 1988.
Carey, Daniel. "The Problem of Credibility in Early Modern Travel." *Renaissance Studies* 33, no. 4 (2019): 524–47.
Cottegnies, Line. "A 'Science of Dreams': 'the Fantastic Ethnography' of Sir Walter Ralegh and Baconian Experimentalism." In *A Knight's Legacy: Mandeville and Mandevillian Lore in Early Modern England*, edited by Ladan Niayesh, 108–27. Manchester: Manchester University Press, 2011.
Daston, Lorraine, and Katharine Park. *Wonders and the Order of Nature, 1150–1750*. New York: Zone Books, 1998.
Donegan, Kathleen. *Seasons of Misery: Catastrophe and Colonial Settlement in Early America*. Philadelphia: University of Pennsylvania Press, 2014.
Edelmayer, Friedrich. "The 'Leyenda Negra' and the Circulation of Anti-Catholic and Anti-Spanish Prejudices." *European History Online (EGO)*, June 29, 2011. https://www.ieg-ego.eu/edelmayerf-2010-en.

Esler, Anthony. *The Aspiring Mind of the Elizabethan Younger Generation*. Durham: Duke University Press, 1966.
Fielder, Leslie A. "Eros and Thanatos: Old Age in Love." In *Ageing, Death, and the Completion of Being*, edited by David D. Van Tassel, 235–54. Philadelphia: University of Pennsylvania Press, 1979.
Fuller, Mary C. "Ralegh's Fugitive Gold: Reference and Deferral in The Discoverie of Guiana." *Representations*, no. 33 (1991): 42–64.
Gallay, Alan. *Walter Ralegh: Architect of Empire*. New York: Basic Books, 2019.
Gilbert, Creighton. "When Did a Man in the Renaissance Grow Old?" *Studies in the Renaissance* 14 (1967): 7–32.
Greenblatt, Stephen. *Marvelous Possessions: The Wonder of the New World*. Chicago, IL: University of Chicago Press, 1991.
Hadfield, Andrew. *Literature, Travel, and Colonial Writing in the English Renaissance, 1545-1625*. Oxford: Oxford University Press, 1998.
Johnson, Amanda Louise. "Nobody's Gold." *Early American Literature* 56, no. 3 (2021): 699–730.
Lefranc, Pierre. *Sir Walter Ralegh Écrivain: L'œuvre et Les Idées*. Paris: Librairie Armand Colin, 1968.
Leitch, Stephanie. "Copies with Wings: Bridging Distances by Printing the Familiar in the Travel Accounts of Theodore de Bry and Levinus Hulsius." In *Far from the Truth: Distance, Information, and Credibility in the Early Modern World*, edited by Michael van Groesen and Johannes Müller, 69–105. London: Routledge, 2024.
Lorimer, Joyce, "Introduction" and "Appendices." In *Sir Walter Ralegh's Discoverie of Guiana*, edited by Joyce Lorimer, xvii-xcvii and 245–331. London: Ashgate, 2006.
Maltby, William S. *The Black Legend in England: The Development of Anti-Spanish Sentiment, 1558-1660*. Durham, NC: Duke University Press, 1971.
Martin, Christopher. *Constituting Old Age in Early Modern English Literature, from Queen Elizabeth to "King Lear."* Boston: University of Massachusetts Press, 2012.
Montrose, Louis. "The Work of Gender in the Discourse of Discovery." In *New World Encounters*, edited by Stephen Greenblatt, 177–217. Berkeley: University of California Press, 1993.
Nicholls, Mark, and Penry Williams. "Ralegh [Raleigh], Sir Walter (1554–1618), Courtier, Explorer, and Author." In *Oxford Dictionary of National Biography*. Oxford University Press, 2022. https://doi.org/10.1093/ref:odnb/23039.
Oldenburg, Scott. "Headless in America: The Imperial Logic of Acephalism." In *The Mysterious and the Foreign in Early Modern England*, edited by Helen Ostovich, Mary V. Silcox, and Graham Roebuck, 39–57. Newark, NJ: University of Delaware Press, 2008.
Oxford English Dictionary. "*Attempt, n., Sense 3.a.*" Oxford University Press, July 2023. Oxford English Dictionary. https://doi.org/10.1093/OED/2199904486.
Riehl, Anna. "Persuading the Prince: Raleigh, Keymis, Chapman, and The Second Voyage to Guiana." In *Tudor Court Culture*, edited by Thomas Betteridge and Anna Riehl, 149–66. Selinsgrove, PA: Susquehanna University Press, 2010.
Schmidt, Benjamin. "Introduction: Ralegh's Courteous Conquest." In *The Discovery of Guiana by Sir Walter Ralegh, with Related Documents*. Boston: Bedford/St. Martin's, 2008.
Sell, Jonathan P. A. *Rhetoric and Wonder in English Travel Writing, 1560–1613*. Aldershot: Ashgate Publishing Limited, 2006.

Tawney, R. H., and Eileen Power, eds. *Tudor Economic Documents: Being Select Documents Illustrating the Economic and Social History of Tudor England.* Vol. 1. 3 vols. London: Longmans, Green & Co., 1951.

Thomas, Keith. "Age and Authority in Early Modern England." In *Proceedings of the British Academy 62*, 205–48. Oxford: Oxford University Press, 1976.

Van Groesen, Michiel and Johannes Müller (eds.). *Far from the Truth: Distance, Information, and Credibility in the Early Modern World.* London: Routledge, 2024.

Vaughan, Alden T. "Sir Walter Ralegh's Indian Interpreters, 1584–1618." *The William and Mary Quarterly* 59, no. 2 (2002): 341–76.

Whitehead, Neal L. "Introduction." In *The Discoverie of the Large, Rich, and Bewtiful Empyre of Guiana, by Sir Walter Ralegh*, 1–116. Manchester: Manchester University Press, 1997a.

Whitehead, Neal L. "Monstrosity and Marvel: Symbolic Convergence and Mimetic Elaboration in Trans-Cultural Representation: An Anthropological Reading of Ralegh's Discoverie…" *Studies in Travel Writing* 1, no. 1 (1997b): 72–95.

4

WONDERS TRAVELLING FROM CHINA

Examples from the Settala collection

Dinu Luca

Introduction: An Elephant and a Book

The list of gifts the ambassadors of Portuguese king Manuel I presented to Pope Leo X in March 1514 was spectacular. Among others, it included a "Moor"-ridden, "Saracen"-guided, well-trained, playful, chaos-inducing, and—emotionally and linguistically—almost human-like elephant; a horse-mounted tamed cheetah wrapped in ermine-lined brocade; cagefuls of Indian fowl, parrots, and small dogs; gems, jewelry, and pearls that "no eye could sufficiently admire and no language could praise enough"; and a silver-made castle-like structure whose central fortress contained an exquisite golden tabernacle and whose many towers housed chests filled with brocade-made pearl-adorned paraments and liturgical vestments heavy with decorations "the most beautiful thing in the world."[1] A richly ornamented pontifical miter joined these pieces of opulent papal attire. Spices and porcelains also changed hands.[2] More significantly, from the perspective of this chapter, another object travelled to Rome on the same occasion—a Chinese book.

The historian Paolo Giovio (1483–1552) does not bring it up when he briefly describes the embassy's splendors in his *Histories of His Own Times*.[3] Instead, he refers to this volume as he digresses on faraway Cathay elsewhere in his text. The city of Canton is like Venice, we learn. The houses there have windows fashioned like "ours." The local means of transportation are also like "ours." The nobler among the Cantonese even resemble Venetian gentlemen in hair-style and gait. Against this background of similarity, Giovio finds "most surprising" only one thing: they have craftsmen who print books the way "we" do. Their tomes deal with histories and ceremonies and consist of long leaves folded in four on the inside. "Of this manner of books," continues Giovio,

DOI: 10.4324/9781003506928-4

"Pope Leo kindly showed me a volume, which had been sent to him as a gift, together with an elephant, by the King of Portugal."[4] Print, concludes our historian, must have reached Europe from China.

Giovio's remarks point to the many challenges that travelling Chinese books must have posed to Europeans. *Air de famille* aside, these printed materials appeared manifestly different from their European counterparts in terms of make-up, craftsmanship, size, weight, textual and visual conventions, imagery, and written character. They tantalizingly asserted the presence of an independent, well-developed cultural, and technological tradition on the other side of the world, capable of codifying unfamiliar meanings in different ways and involving distinct materiality. On the other hand, Chinese books were so rare that physically interacting with them was a privilege. For people like Giovio, they must have been at the extreme of wonder: after all, the historian writes of a Chinese tome accompanied by an elephant—that paragon of "unusualness and inconceivability"—not vice-versa.[5]

This essay is about another Chinese volume and two pages from another one. As far as we know, no pachyderm escorted them into the celebrated gallery of Manfredo Settala (1600–1680), one of Europe's paramount collectors of the early modern age, more than a century after Dom Manuel's book reached the Pontiff's hands. Nevertheless, even at such a late moment—many decades after the Christian missions had gained solid ground in China and increased interaction had made Chinese objects (including books) noticeably more common in Europe—they were still perceived as travelling wonders. According to what we know about Settala's collection, they did not belong in Manfredo's impressive library. Physically, they found their place next to myriad non-textual *mirabilia* on a crammed bookshelf in one of the collection's rooms; categorially, they joined similarly marvelous objects in a set of exotica whose very existence testifies to the perpetuation of Giovio-like sensibilities well into the 17th century. That does not mean that, when dealing with a richly documented context from so long after the early contacts that had enabled the pilgrimage of Dom Manuel's Chinese tome, we are not in a better position to learn more about such materials. The present contribution aims to show this.

I proceed in three steps. First, I provide a quick overview of Chinese books' material and textual presence in Europe by the 1660s, focusing on a particularly influential account and its framing in the Italian versions of the Settala collection catalog. Next, I briefly introduce Manfredo's gallery, zooming in on its Chinese paper-based materials as discursively and visually represented in contemporary sources. Lastly, I concentrate on two items, reframing previous scholarly findings in the case of the former (two pages from a daily-use Chinese encyclopedia) and identifying the latter (a pulse-medicine text) and hypothesizing on its travels from a Chinese printshop into the hands of a famous Milanese collector. I conclude by emphasizing the multifaceted significance of Settala's

Chinese objects selected for discussion and the importance of persistently interrogating them and similar items.

Chinese books in Europe, ca. 1660

Dom Manuel's book was not the only Chinese tome Giovio saw. In 1548, the historian João de Barros sent him "*um livro da escritura dos chis*,"[6] which may have joined other wonders (like Prester John's shirt or Indian nuts) in Giovio's collection in Como.[7] Barros could afford to make such an extravagant gift: he had several Chinese volumes (including an atlas and a "courier manual") in his possession.[8] While unusual, such extensive ownership of Chinese printed materials in mid-16th-century Europe points to an ever-growing Europe-bound traffic in knowledge and wonder for the benefit of a public whose appetite for things from and words about the East was becoming insatiable.

The following decades bring new proof in this sense. Passing mentions multiply: Montaigne famously depicts the delicate materiality of a Vatican-held Chinese tome written in a "*charactere sauvage*";[9] others signal the presence of Chinese books at princely courts or in the hands of those connected to the ever-expanding missionary networks;[10] yet others grow entranced with the Chinese script, describing, analyzing, and categorizing it, and working to see it in European print. In time, potentate collections become richer;[11] university and private scholarly libraries begin displaying Chinese items;[12] more and more Chinese books—often poorly printed cheap copies and disparate fascicles—as well as all sorts of other blank, handwritten, or printed paper materials reach Europe, where they trigger unabated interest as they are sold, auctioned, exchanged, gifted, refashioned *à l'européenne*, collected by everybody from polymaths to merchants and cardinals, and subjected to countless attempts of deciphering.[13]

They are also discoursively disciplined. One of Europe's earliest works dedicated explicitly to China already includes a chapter on Chinese writing in 1577;[14] not much later, another one describes the many Chinese books mined for information by inquisitive missionary minds;[15] learned disquisitions on print, languages, or script feature in various efforts to aggregate new knowledge into comprehensive syntheses.[16] In the next century, China master texts further refined matters, articulating authoritative accounts whose disseminative power was matched only by stylistic triumphs like Daniello Bartoli's 1663 *Cina*. It is from two texts in the former category, the Matteo Ricci–Nicolas Trigault 1615 *De Christiana expeditione apud Sinas* and Álvaro Semedo's 1642 *Imperio de la China*, that physician Paolo Maria Terzago—the compiler of the first published catalog of Settala's collection—picks (via collector Ole Worm's 1655 *Museum Wormianum*) a few points to "reveal the wonders of Chinese writing to the reader."[17] It is from the latter compilation that another physician, Pietro Francesco Scarabelli, the author of the engaging 1666 Italian rendition-cum-rewriting of Terzago's 1664 sober Latin work, makes a much more extensive selection.[18] Given that Bartoli's remarks can also be seen as offering an accurate (albeit insufficiently detailed) image of what an inquisitive European was likely to know of Chinese books and associated matters one and a half centuries after an elephant accompanied a Chinese tome from Lisbon to Rome, a brief review of these perspectives as embedded in Scarabelli's catalog is now in order.

At the beginning of his extended digression, the physician explains that he intends his Bartoli selection to satisfy the readers' curiosity about the "truly singular" ways the Chinese speak and write. He starts with Bartoli's statements on these matters, constantly pendulating (like the Jesuit source he faithfully replicates) between writing and speech. First, we learn about the non-alphabetic Chinese script and the massive number of "ciphers" its users must master. Luckily, the characters share some common ground, which, according to Bartoli, derives from the Chinese phonetic specificity. After a brief detour examining the invention of writing in China, Bartoli continues with a short paean to this script, which is deservedly admired by those who notice "the most intriguing delineation" of its thousands of ciphers, "understand its art and mysteries," and realize it "is not all play of quills moved by whim, but rather, in large part, invention and labor of ingenuity, perhaps one with even more spirit than that of Egyptian hieroglyphics." Unfortunately, speech is not on par and engenders ambiguity and misunderstanding. Once again, writing can solve such issues because "what is mysterious" in Chinese characters make them appropriate for universal communication everywhere in China (regardless of dialectal differences) and in East Asia at large. Mystery, however, does not stop here. Next to topolects and the standard, "forensic" language, the Chinese have a literary variety, which is "highly capable of art and wisdom: since their characters have the force of hieroglyphics, to know how to combine them and make them into compositions of beautiful mystery is a game of great ingenuity." The somewhat bland observations on Chinese grammar that follow fail to dilute the pervading sense of wonder Bartoli instills throughout. As he insists in a remark that Scarabelli leaves out, Chinese has thousands of other "notable properties" that he will discuss elsewhere.[19]

Scarabelli's selection from Bartoli's subsequent considerations is less extended. The physician copies the Jesuit father's points on the Chinese brush-based penmanship (which involves "a peculiar manner most unsuitable to us, [but] to them… admirably suited"), sense and direction of writing, and the preparation of liquid ink from solid ink cakes. He bypasses, however, Bartoli's long passages on Chinese print (technology, varieties, price, speed, and illustrations) and personal ownership of "works on large sheets… [showing] human figures, animals, trees, flowers, and certain oddities of their whimsy." Bartoli's final comments on Chinese paper, emphasizing its frailty, delicacy, and uncertain origin, also end Scarabelli's lengthy excursus.[20]

Bartoli's synthesis testifies to the enduring European sense of wonder about Chinese books—and, indeed, all things Chinese—that intensified contact and the accompanying increase in materiality steadily fueled: if anything, familiarity bred admiration, and mystery was everywhere. Scarabelli's decision to reproduce Bartoli's remarks instead of translating Terzago's (and Worm's) short, restrained passage points in the same direction. Moreover, Scarabelli's choice to quote extensively from a book on China rather than other realms newly opened to European investigation suggests a hierarchy of wonder that is substantiated by both the contents of Settala's book-related exotica (dominated by Chinese items) and the ways the materials documenting Manfredo's cabinet verbally and visually depict them. Before exploring these aspects at some length, we need to learn more about this celebrated Milanese collector, his "museum or gallery," and the words and images we have about it.

Manfredo Settala, his collection, and his Chinese book-related items

Manfredo Settala was no ordinary person.[21] He was well-born: his more remote ancestry included beati, bishops, and major civic figures; his father, Ludovico, a philosopher and Milan's chief medical official, gained extraordinary distinction (and also literary immortality under the pen of Alessandro Manzoni) fighting the 1629–1631 plague. He was well-educated: after early studies at the Jesuit college in Brera (a period when he chanced to see "the rarest wonders" housed in Mantua's Palazzo Ducale), he continued in Pavia, Siena, and Pisa, obtaining a law degree, acquiring Medici patronage, and making friends with a future pope. He was well-travelled, with years adventurously spent at sea and first-hand knowledge of the Eastern Mediterranean. He was also well-connected. From an early age, he joined the networks of influential Cardinal Federico Borromeo, who also offered him a church position that provided him with revenue and leisure. As the prestige of his collection increased, his gallery welcomed princes, ambassadors, and sundry potentates, many of whom gifted him rare and valuable objects. He was in touch with numerous scholarly luminaries in person or through letters. He also hosted countless visitors, some sketchily or substantively narrating their experience in his museum.[22] Lastly, he was well-respected: dithyrambs accumulated in his lifetime, his funerals occasioned multiple, dizzyingly intricate Baroque spectacles (in church, in the streets, and in print), and one full-fledged biography appeared two years after his death.

Manfredo was also a maker and a showman. When he inherited his father's library and collection stump, he had already lathe-produced many elaborate, delicate "bagatelles" like the tiny piece of turnery he melancholically holds in a 1622 portrait.[23] Mirrors, telescopes, musical instruments, quasi-perpetuum mobiles, porcelain-like pieces, automata, and myriad objects will follow. "The Milanese Archimedes" used these and other contraptions to amuse and amaze the many guests flocking to the several rooms housing what will ultimately become a perhaps 5,000-object-strong collection and a 10,000-volume library. However, his showmanship extended further than that. Manfredo performed his "supremely witty text" of a collection for the benefit of his callers, making things happen in front of their eyes through experiments and play but also sucking them into its "abyss" through contextualizing, narrativizing, and interpreting.[24] Such multifaceted embedding did more than impart knowledge and gain prestige for the collection's owner, not least because it could involve reciprocity—like when well-travelled Jesuit fathers illuminated some of the many mysteries of Settala's objects in well-staged spectacles of superb erudition. On the other hand, this whole dynamics of producing wonder by making, making happen, and making sense breathed even more life into Manfredo's polyphonic, metamorphic collection.

Much missed after Settala's passing, such vividness of the museum experience perhaps best transpires from two works documenting his gallery. The first is a manuscript "Indice" of the curiosities in Manfredo's house, which offers a space- rather than a category-centered picture of his cabinet (as the catalogs provide).[25] At least to some degree, browsing this text mimics the dynamic experimentation of the museum space. Of course, this must have been infinitely more immediate, embodied, and highly sensorial, but also selective and context-dependent, than whatever the "Indice," despite its cumulative narrativization, could ever afford.[26] Nevertheless, the image we form of these cluttered, chaotic, and wonder-filled rooms remains enticing many centuries after their contents had been dispersed, and the myriad conversations developing within these "quasi-inextricable mixtures" of *naturalia* and *artificialia* had long become silent.[27] While this is not the place to explore such multiple dialogues, a few remarks about the Chinese books in Settala's collection and their context at the moment the "Indice" coalesced are in order.[28]

The "Indice" contains fewer entries about exotic book-related items than the printed catalogs. It lists two books about heaven written in "very beautiful characters and figures," some large-size blank Chinese paper, other sheets of Chinese and Japanese paper, and a man-sized print depicting heaven and hell through characters and figures.[29] These objects and "Three Arabic books excellently written in figures" shared the same shelves with an infinite variety of pieces—among others, "Oriental mother-of-pearl snails," miniature Amsterdam-made globes, clocks, balsam-holding vases, precious wood and woodwork built from it, asbestos-made objects, China-produced combs, eaglestones, glossopetrae, lapis-lazuli, "a root found in Grand Cairo," knives, malachite, beryl, stuffed desmans, byssus-made cords, gold-decorated Persian bags, and exquisite "Indian" fabric.[30] In this corner of the much broader landscape of wonder depicted by the "Indice," Chinese books belonged just as naturally as a lonely volume did with Dom Manuel's menagerie.

The second work I alluded to above—i.e., a set of drawings authored by young Milanese painters selectively documenting the collection—also captures the intensity of the visitors' encounter with the Settala gallery. Not only do Manfredo's wonders come back to life, in vivid colors and with much detail, in images displaying nothing of the "great vagueness" Scarabelli writes about, but they also do so accompanied by their owner's own words.[31] Indeed, most of the 300-odd drawings in existence carry a short caption in Settala's hand. Going through the five surviving notebooks thus resembles a guided visit in which the immediacy of a textualized voice concisely or exuberantly identifying, comparing, sensorializing, biographizing, tracing provenance (and strengthening prestige in the process), or indulging in brief scholarly pyrotechnics suits well the "rarities" chosen for visual reproduction (and discoursive exposition).[32] To what degree such a metaphorical visit corresponds to the physical experience of the "wild, exuberant, encyclopedic inclusiveness" of Manfredo's collection is, however, debatable.[33] While the categories into which the image groupings in the notebooks indirectly discipline the collection are less well-circumscribed than those in the catalogs, an itinerary among objects matching the progressive intimacy permitted by the "Indice" is also absent. A few words about the Chinese book-related materials represented in the drawings are now in order.

These include the large sheet of blank paper mentioned in the "Indice" (Figure 4.1, top), handwritten documents (Figure 4.2), and one of the two Chinese pieces the present contribution concentrates on (see below).[34] Their images share the four notebook folios dedicated to exotic writing and its materiality with visualizations of two Arabic manuscripts (Figure 4.1, bottom, and Figure 4.8) and a "Tamolese or Malabar book made of tree bark or leaves written on with an iron stylus" (Figure 4.5, top). While very different from what the "Indice" placed in the immediate company of such paper-based exotica, the objects depicted on the manuscript's surrounding folios—among others, a Japanese box, a Turkish ink-slab, ritual garb from Brazil, Peruvian feather-made decorations, "blowflies from Congo," South American hammocks, a Chilean broom, African carpets, exotic shoes and boots, bows, arrows, quivers, crossbows, knives (and a fork), balms, Chinese stones and rhubarb, nutmeg and other fruits, exotic beans and bean pods, tree bark, bezoars, horns, fossilized wood, and a mandragora root—attest to the same overwhelming regime of wonder Settala associated with the textual-scriptural materiality of the Other.

When seen against the accumulated wondrousness built by the "Indice" and the notebooks, the categorial textualization of Settala's museum in the printed catalogs appears significantly less exuberant, not only in Terzago's work—whose otherwise "tersa Latinità" is noticed by Scarabelli—but even in the latter's expanded Italian versions. While the topicality of the sixty-eight chapters making up the catalogs can feel occasionally disorienting, a sense of order and

FIGURE 4.1 Biblioteca Estense Universitaria, ms. Campori γ.h.01.21, 13r. Su concessione del Ministero della Cultura - Gallerie Estensi, Biblioteca Estense Universitaria.

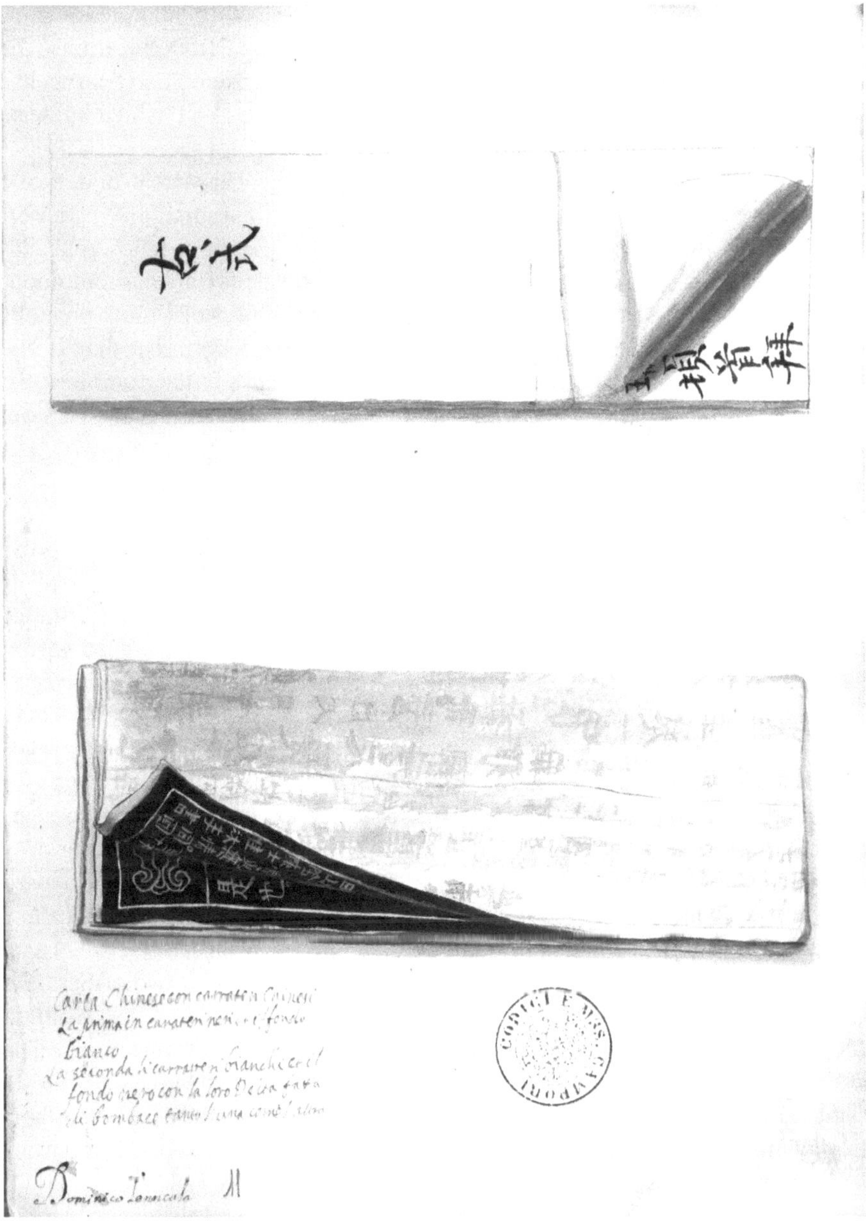

FIGURE 4.2 Biblioteca Estense Universitaria, ms. Campori γ.h.01.21, 11r. Su concessione del Ministero della Cultura - Gallerie Estensi, Biblioteca Estense Universitaria.

structure prevails. The indexes, the numbering, and the marginal notes all contribute in this sense; so do the shorter or longer scholarly explorations, the learned asides, or the quotations liberally included in Terzago's and Scarabelli's "models of wit and persuasion."[35] A brief look at chapter LVI, which interests us here, will further illustrate these points.

As its respective titles clarify—"Books in strange characters and exotic paper also made from leaves or tree bark" in Terzago's original and "On peregrine books, foreign paper, made from leaves and tree bark" in Scarabelli's reworkings—this section treats of objects that find comfortable accommodation next to balms, oils, knives, "various and peregrine crafts of India," precious chests, and glass globes (the topics of surrounding chapters) rather than in the Biblioteca Settaliana described in chapter LXV. Terzago's initial numbered list includes sixteen entries, which grow to nineteen in Scarabelli's 1677 version; with multiple items under one entry and hazy phrasing ("diverse alte carte"), it is impossible to establish the exact number of objects.[36] Like in the "Indice" and Settala's handwritten captions, information remains sparse (except when visual elements permit ekphrastic episodes) and concerns mostly materiality and provenance. The floating referent status of labels such as "Chinese," "Japanese," or "Indian" further complicates matters.[37] Notwithstanding the objects' resistance to detailed verbalization (the two physicians would have surely agreed with a visitor's remark that such "exotick rarities… are better seen then described"),[38] an interest in spectacle at the expense of precision is also noticeable. Both authors showcase personal scholarly competence by appealing to master texts. They also delight—exuberant Scarabelli more than reserved Terzago—in depicting deciphering tours de force performed on some of Manfredo's objects by travelling Jesuit guests. Tellingly, however, they refrain from writing anything specific (title, detailed contents) about the Chinese book subjected to Father Michał Boym's decrypting (see below) or the Malabarese text that thoroughly captivated another visiting father, Giacinto de Magistris.

We need not follow the same path. Below, I concentrate on two of Settala's Chinese objects. The former is an item detailedly discussed by Arianna Magnani, which I further frame in the context of Manfredo's collection.[39] The latter is a book that painter Domenico Tencalla drew for Manfredo's notebooks (Figure 4.4). I identify the volume, discuss the fate of the text during its European travels, and speculate on how Settala may have acquired his copy.

Two pages and one book: conjectures and certainties

Scarabelli introduces the item listed under number 10 in chapter LVI as "Various other sheets made of the thinnest silk paper showing strange Chinese garments and birds." The Ambrosian (S.Q.V.VIII.15.) holds two non-sequential pages (now glued together, with the left one in reverse) corresponding to this description. These can be traced to one of the many extremely popular,

Fujian-province-cheaply-printed, daily-use encyclopedias of late Ming times, specifically to the section typically entitled "Various barbarians" 諸夷 that such works dedicated to the comprehensive, complexly articulated spectrum of the Other in Ming minds.[40] The two-register structure that commonly builds the pages of such chapters (Figure 4.3) here gives place to a ternary arrangement whereby a bottom text-only register refashions the usual text-to-image balance.[41] At the top, we find textual and visual depictions of mythical beings (on the Ambrosian pages, birds) featured in the early Chinese cosmography-cum-bestiary *The Book of Mountains and Seas* (*Shanhaijing* 山海經). In the middle, the excerpt from the encyclopedia-geared reworking of *The Record of Naked Creatures* (*Luochonglu* 臝蟲錄)—a heavily illustrated text on exotic (human-like) people and their realms—identifies, briefly describes, and offers the visual likeness of four such exemplars of otherness.[42] Lastly, the lower register text

FIGURE 4.3 Pages from Liu Ziming 劉子明, ed., *Xinban zengbu tianxia shiyong wenlin miaojin wanbao quanshu* 新板增補天下便用文林妙錦萬寶全書 [Complete Book of Myriad Treasures: Fine Brocade from the Forest of Letters for Use in the World. Newly Carved, Expanded, and Supplemented]. Harvard-Yenching Library, Harvard College Library, Harvard University, T 9299/7224.

differs between the two Ambrosian pages. The one on the right continues a passage on the previous page and lists so-called alien rarities, like sandalwood, crystals, flammable liquid, a Buddha stone, rose water, and other desirable wonders from "Champa."[43] The one on the left finishes the description of a Central Asian other initiated on the preceding page.[44]

The two sheets have more to say. Specificities in setup, phrasing, and imagery interestingly set them apart from their peers in similar chapters.[45] While this is unsurprising, given what we know about such texts' production context and circulation, the presence of the passage on mysterious Central Asians (who seem to be textually stitched from bits of anthropological lore on two different peoples)[46] is intriguing. Hypothesizing about this and other issues raised by these two lonely, Europe-based survivors of a long-lost Chinese encyclopedia is beyond this essay's scope. On the other hand, their inclusion among Manfredo's collectibles is worth some discussion.

We should note, first, that the Ambrosian-housed glued items were part of a larger group (Terzago speaks of "many" rather than Scarabelli's "diverse" leaves) of connected pages. During its travels, the original volume must have become a kind of super-semiophore, i.e., something that could be disconnected from its initial multi-fascicle encyclopedia setting and further decontextualized by mutilation at no loss. The presence of imagery placing the pages into an order of wonder both exotic and enticingly familiar to European eyes—it is worth pointing out that in the *Luochonglu*-inspired register, one can occasionally find represented, next to Koreans and Japanese, dogheads, winged people, or giants—would have rendered the cohesion of the whole even less significant. The equalization of wonder in the space of the museum, intimated by all our sources, confirms this—every wonder-imbued object deserved collecting, with no differences in prominence between fragments (separate sheets) and books, illustrated and non-illustrated materials, or written and unwritten objects. Such equalization is, however, offset by a discursive order that privileges Chinese book-related elements and therefore suggests—as we have seen—distinct hierarchies, at least in the textual regime of wonder.

The catalogs refer to the figures in the middle register as wearing "Chinese garments." On the one hand, this seems to imply that the leaves Settala owned did not include images of the less-than-human peoples sometimes featured in the *Luochonglu*. On the other, it makes an equation abundantly clear: what for the Chinese illustrator belonged to the realm of the other, for the European observer was "Chinese"—a descriptor well on its way to becoming Europe's favorite shorthand term for "wonder." This should not prevent us from perceiving the museumification at work—of visual and textual knowledge into the "theater of marvels" of a Ming dynasty daily-use encyclopedia (more specifically, here, into the "textual cabinet of curiosities" filled with novelty and rarity of the *Luochonglu*) and of two travelling pages carrying such knowledge within the space of a 17th-century Milanese gallery.[47] The impossible imbrications suggested by all this—a chapter containing "the whole world" becoming part of "the whole world" contained in Manfredo's collection—would have made Settala, the producer of countless Baroque contraptions and consummate mirror maker, smile.

One last point could be made before we move on to the second object of interest here. The catalogs include, under number 1 in the list, not one but three pieces, the second of which Scarabelli describes as a "Chinese book… all full of maps of cosmography with [people wearing] Chinese garments." Could this item, which our Milanese searcher of the other probably already owned when the "Indice" was composed, also have been part of a daily-use encyclopedia (looking perhaps like in Figure 4.4)? Such a volume would not have invited quick dismemberment. It would have also fit better with the other materials under Scarabelli's number 1 (a sumptuous cosmography and a Jesuit-made map) than the pages filled with other "Chinese people" shown in the company of strange birds under number 10. While this is speculative, the significant presence of such encyclopedias in European libraries raises such a possibility.[48]

We can have more certainty about another object in Manfredo's gallery (Figure 4.5, bottom). Like with the other written Chinese materials drawn by Tencalla for Settala's notebooks, the characters' legibility is striking. A comparative look at an image in the printed catalog of Ludovico Moscardo's contemporary collection makes this abundantly evident (Figure 4.6). Moreover, the young artist chose to represent everything he saw, i.e., a half-open book and pages with complete strings of characters. He also preserved the page layout, included the fascicle number of the work he reproduced, and paid attention to character size (thereby informing us that the text was accompanied by commentary). All this permits the accurate identification of the book once touched by Manfredo's, his painter's, and his visitors' hands in Milan sometime in the mid-17th century.

The volume in question is the third (and last) in a three-fascicle commented edition of a pulse-related text in the complex Wang Shuhe *Maijue* 王叔和脈訣 tradition (more on this below).[49] While Tencalla misses characters and truncates strings, his accuracy is such that even the least visible row on the right page (which the book fold somewhat obscures) contains at least some recognizable characters. The same is true for the barely visible corresponding left row, where, with some luck, one can identify the bottom last characters as 手 (hand) and 足 (foot) (Figure 4.7, left).

Settala's caption contradicts this identification. He notes: "The other Japanese book that treats of the virtues of plants, as I have been told by the Ambassador of China, who came with Father Bovino, who read it for more than two hours." An entry in the printed catalogs (under number 2) offers more details about this work and the deciphering episode. This is what Scarabelli contortedly writes: "Chinese book of paper similarly made of the thinnest silk in which are written and described the faculties and the stupendous virtues of

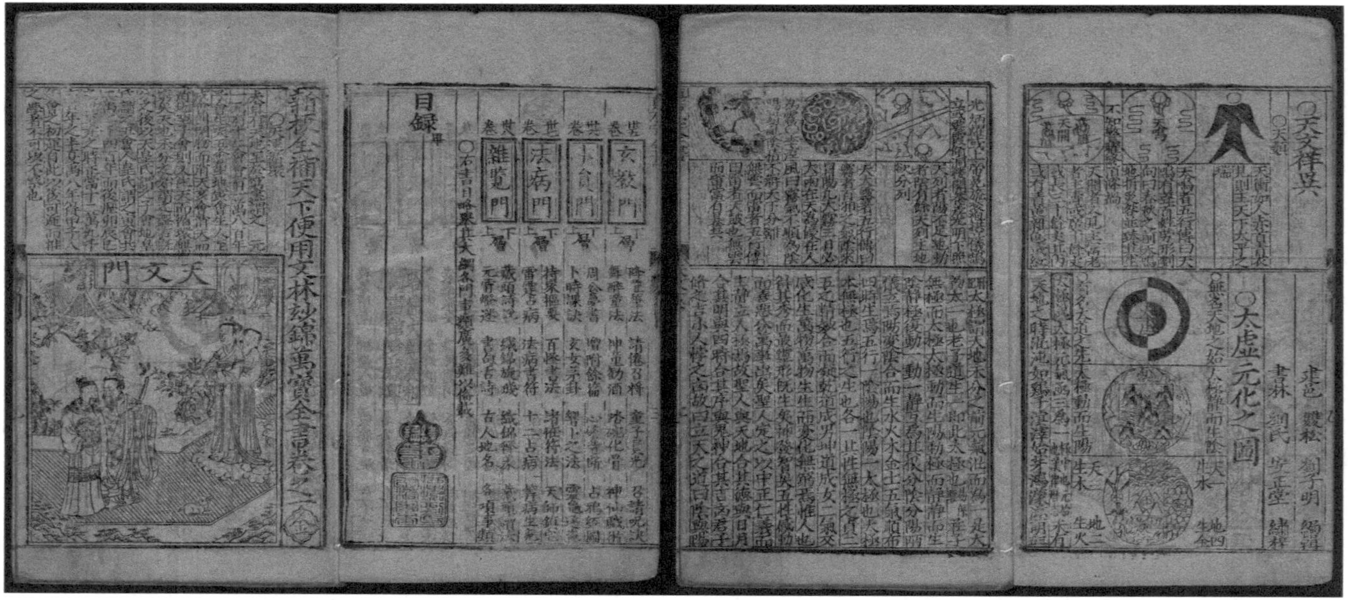

FIGURE 4.4 Pages from Liu Ziming 劉子明, ed., *Xinban zengbu tianxia shiyong wenlin miaojin wanbao quanshu* 新板增補天下便用文林妙錦萬寶全書 [Complete Book of Myriad Treasures: Fine Brocade from the Forest of Letters for Use in the World. Newly Carved, Expanded, and Supplemented]. Harvard-Yenching Library, Harvard College Library, Harvard University, T 9299/7224.

FIGURE 4.5 Biblioteca Estense Universitaria, ms. Campori γ.h.01.21, 14r. Su concessione del Ministero della Cultura - Gallerie Estensi, Biblioteca Estense Universitaria.

FIGURE 4.6 Drawing in *Note overo memorie del Museo di Lodovico Moscardo…*, 125.

the plants of those countries, as is attested by the inscription one can read in Portuguese characters on the cardboard in addition to what added in confirmation Fr. Bohino of the Company of Jesus, who years ago [when he] came to honor this Museum, together with the Legate of the Queen of China, read it with singular satisfaction and to the unspeakable pleasure of those present, he being very expert in that idiom, and when entreated by our Signor Manfredo to write something in Chinese, he delighted him with a particular inscription, which is also preserved here."[50]

Scarabelli's entry establishes (or suggests) identities and contexts more clearly. The mysterious ambassador—"the Legate of the Queen of China"—is doubtlessly Andreas Zheng, a young Chinese man who accompanied Polish Jesuit father Michał Boym (Father "Bovino" or Scarabelli's "Bohino") on a European mission on behalf of the last Southern Ming emperor resisting the Qing dynasty takeover.[51] Boym's exuberant performance may have happened in the winter of 1655–1656, when the two travellers on their way to China via Genoa may have stopped in Milan. The book giving such unspeakable elation to the Jesuit is indeed Chinese. However, this is not the same volume that

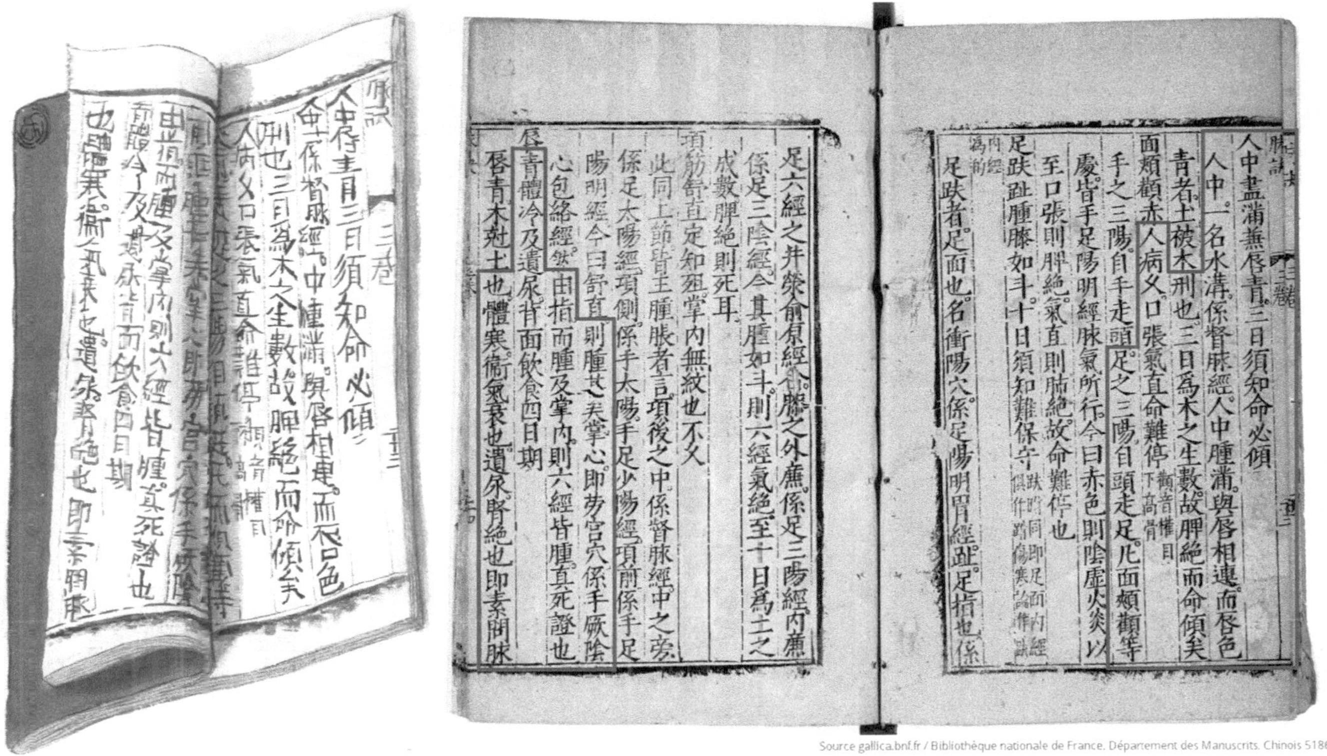

FIGURE 4.7 Left: Tencalla's drawing. Biblioteca Estense Universitaria, ms. Campori γ.h.01.21, 14r, bottom. Su concessione del Ministero della Cultura - Gallerie Estensi, Biblioteca Estense Universitaria. Right: The page Tencalla reproduced, with his selection highlighted. Bibliothèque nationale de France, BnF Chinois 5186, fols. 23v–24r.

Tencalla's drawing shows. Boym—who, incidentally, rendered into Latin a text in the *Maijue* tradition and would have recognized Settala's exemplar for what it was[52]—must have held in his hands a herbology text, which Manfredo (when annotating Tencalla's drawing) and hence Terzago and Scarabelli mistook for what the painter so accurately represented.

The full name of this particular work is *The Corrected Interpretation of the Maijue with Commentaries and Proofs by Ma Xuantai* 刻馬玄臺先生註證脈訣正義, authored by a little-known Ma Shi 馬蒔 sometime in the second half of the 16th century.[53] It offers a commentary on a popular, rhymed text on pulse diagnosis, which circulated widely and in many editions. While no copy exists in China or Japan, the situation differs in Europe. The BnF holds two fascicles (2 and 3) of this text, with the third (whose fols. 23v–24r are those reproduced by Tencalla) being divided between mss. Chinois 5187 and 5186. The Bodleian holds all three fascicles (Sinica 11/1 and 11/2), gifted by two different donors. The first and third fascicles are incomplete, with Tencalla's pages being absent. A fragmentary third fascicle, also missing the pages drawn by Tencalla, is preserved in the Herzog August Bibliothek in Wolfenbüttel (Cod. Guelf. 130.4 Extrav. [1]). An incomplete first fascicle has been recently identified in the Utrecht University Library (shelfmark V oct 857).[54] Lastly, the Ambrosian also houses a *Maijue* (S.Q.V.VIII. 15/4), which is held to have reached the Ambrosian in Borromeo times via the scientifically inclined China Jesuit Johann Schreck.[55] Upon examination, this turns out to be an incomplete third fascicle of the same Ma Shi commentary,[56] which, I believe, is what Settala owned, Tencalla drew, and Terzago/Scarabelli (under number 3) and Settala in his caption referred to as "the other Japanese book."

How did Manfredo come by his copy? Had he received it from a prominent donor or a travelling Jesuit scholar, he would have mentioned names and perhaps circumstances. However, a quick comparative look at the "Indice" and the catalogs reveals that this does not happen with every donated object. Alternatively, Settala could have acquired his book from the market—we know he was willing to invest good money in "*libri bell[issi]mi*."[57] With no clear answer seemingly possible, the only certainty we may end up with is that, since there is no entry in the "Indice" about this object, the tome must have entered the collection sometime in the 1650s, like most other items among the exotic paper-related materials. Considering that Boym did not get to effuse over a text he was undoubtedly familiar with during his visit, it appears likely that Settala did not own it in 1656. It would have reached Manfredo without the other fascicles like countless other Chinese tomes coming into European collectors' hands. This divisibility of wonder should not prevent us from noticing the truly wonderful thing behind a book's travel from a Chinese printshop to an Italian virtuoso's cabinet: its survival. The same, of course, is true about the two encyclopedia pages discussed above. Through their rarity and the mysteries they still hide—what work do the two lonely sheets belong to?; how did they and the pulse diagnosis text travel to us?—both items are now scholarly semiophores, just as valuable to us as they were to Manfredo when he showed them to his gallery's visitors.

Conclusion

The catalogs and the drawings can tell us more about Manfredo's treatment of book-related wonder. Objects coming from a closer other are identifiable: this is the case with item 14 in the catalogs, declared by knowledgeable people (says Scarabelli) to be the Quran. Tencalla agrees and draws two pages showing Versets 14–20 from Sura 88 (Figure 4.8). While the same cannot be said about the other Arabic manuscript (bought by Settala in Aleppo, as he asserts in his handwritten caption), the young painter's drawing is precise enough to allow scholarly deciphering today: the text deals with syntactic analysis (Figure 4.1, bottom).[58] This creates an instructive contrast with Manfredo's "Tamolese or Malabarese" book (Figure 4.5, top), which overwhelmingly captivated Jesuit father Giacinto de Magistris and offered him the opportunity to perform even more dazzling deciphering/enciphering spectacles than those involving Boym.[59] As stated above, we learn little from the father's hermeneutic performance—the "many beautiful curiosities hidden in those characters" Scarabelli writes about remain obscure. Moreover, Tencalla's interest lies so much in the object's materiality that we see very little of the writing therein, with characters appearing particularly "peregrine."

This paper has shown that this is not the case with Chinese signs. Everywhere we see them, we can read them—including, through digital manipulation and only partially, even on the back of the page whose thinness Tencalla wants to emphasize by reproducing them as they appear mirrored (Figure 4.2, bottom).[60] This domestication of Chinese characters in European discourse and print—which is perhaps also at work in Terzago's chapter on "Books in strange characters" becoming Scarabelli's section on "Peregrine books"—is undoubtedly ongoing in 1677, when the second (and last) edition of the catalog was published. This is just a few years earlier Athanasius Kircher's famous *China illustrata*... not only reproduced drawings from the calligraphy section of a daily-use encyclopedia similar to Settala's (Figure 4.9) but also included numerous, large-size separate characters as well as a reproduction of a 1,900-character-long famous inscription.

After Kircher's ultimate Baroque triumph, Settala-like wonder about Chinese books and signs could but subside. Chinese tomes steadily joined their European peers on library shelves rather than the extraordinary smorgasbord of *artificialia* and *naturalia* where the "Indice" positions them at one moment in the history of Manfredo's collection. Deciphering efforts increased, translational tours de force were published, and general scholarly disciplining heralded the birth of modern sinology in the early 19th century. As for

FIGURE 4.8 Biblioteca Estense Universitaria, ms. Campori γ.h.01.21, 12r. Su concessione del Ministero della Cultura - Gallerie Estensi, Biblioteca Estense Universitaria.

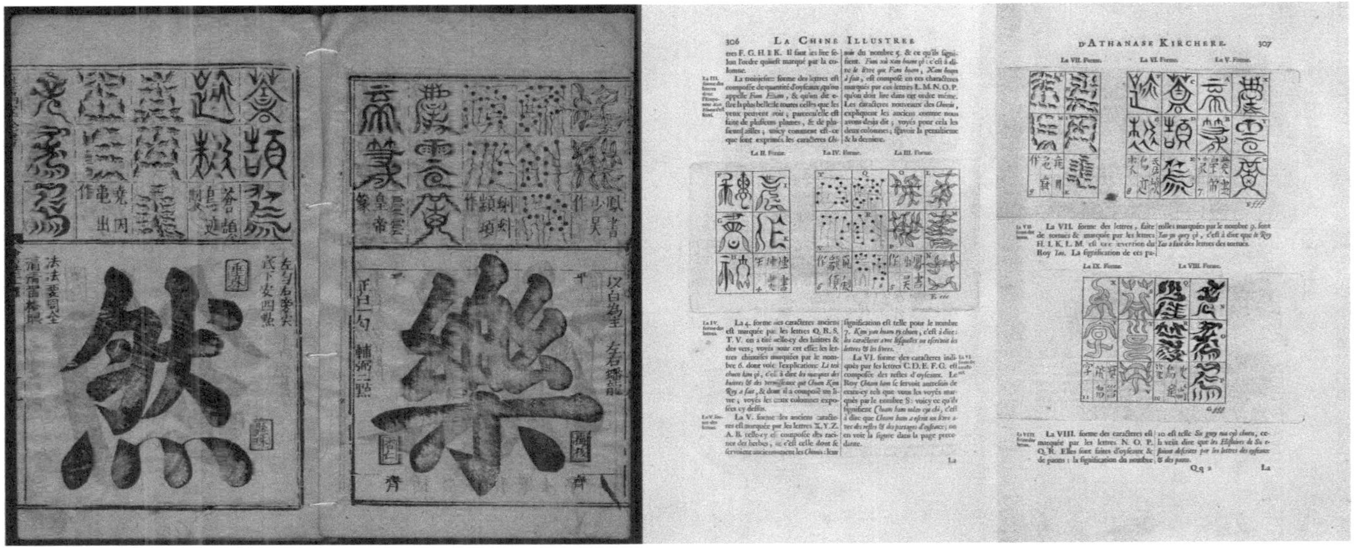

FIGURE 4.9 Left: Pages from Liu Ziming 劉子明, ed., *Xinban zengbu tianxia shiyong wenlin miaojin wanbao quanshu* 新板增補天下便用文林妙錦萬寶全書 [Complete Book of Myriad Treasures: Fine Brocade from the Forest of Letters for Use in the World. Newly Carved, Expanded, and Supplemented]. Harvard-Yenching Library, Harvard College Library, Harvard University, T 9299/7224. Right: Pages from Athanasius Kircher, *La Chine d'Athanase Kirchere…*

Chinese characters, these started leaving books and engaging in their own trajectories. In the last years of Settala's life, their avatars could already be found decorating Delftware.[61] After that, neither 18th-century chinoiserie nor its countless subsequent (sub)versions have failed to incorporate them in their vocabularies of charm or in their attempts to recover wonder in a disenchanted world.

Nevertheless, at a time when few cabinets of curiosities still cherish printed and manuscript wonders travelling from afar, not everything is lost. As this paper has argued, there is still much to learn about such objects and their voyages. While we may never know enough about routes, intermediaries, authors, titles, contents, or dates, shards of information can still be gleaned. In an age when Chinese books and elephants rarely, if ever, travel in each other's company, this should be reason enough to embark on such quests.

Acknowledgments

This work was supported by the National Science and Technology Council, Taiwan [112-2410-H-003-067-MY3]. An early version of this piece was presented at the 2022 ACLA annual meeting. I thank Noël Golvers for his help with information about Michał Boym.

Notes

1 Federico Stefani et al., eds., *I diarii di Marino Sanuto*, 58 vols. (Visentini, 1879–1903), vol. 17 (1886), col. 422; vol. 18 (1887), cols. 58–60; 85–86; Salvatore de Ciutiis, *Une ambassade portugaise à Rome au XVIe siècle* (Michele D'Auria, 1899); Donald F. Lach, *Asia in the Making of Europe*, 3 vols. (University of Chicago Press, 1965–1993), vol. 2.1 (1970), 135–44; Silvio A. Bedini, *The Pope's Elephant* (Penguin Books, 1997).
2 Lach, *Asia*, 2.1, 41; Bedini, *Pope's Elephant*, 27.
3 Paolo Giovio, *Histoires de Paolo Jovio*, trans. Denis Sauvage, 2 vols. (Guillaume Rouille, 1561), 1: 224.
4 Giovio, *Histoires*, 1: 307.
5 Luís Augusto Rebelo da Silva, ed., *Corpo diplomatico portuguez*, vol. 1 (Typographia da Academia Real das Sciencias, 1862), 236.
6 João de Barros, *Ásia*, ed. Hernani Cidade, 4 vols. (Divisão de Publicações e Biblioteca. Agência Geral das Colónias, 1945–1946), 1: 360–61.
7 Sylvie Deswarte-Rosa, "Échos de la Chine à Rome en provenance du Portugal au XVIe siècle: Paolo Giovio, Giovanni Ricci da Montepulciano, Frei Bartolomeu dos Mártires," in *Património cultural chinês em Portugal*, ed. Luís Filipe Barreto and Vítor Serrão (Centro Científico e Cultural de Macau, 2015), 45.
8 Barros, *Ásia*, 1: 353; 3: 91; Liam Matthew Brockey, "The First China Hands: Iberian Visitors to the Ming Empire and the Forgotten Origins of Sinology," in *Western Visions of the Far East in a Transpacific Age (1522–1671)*, ed. Christina Lee (Ashgate, 2012), 75–76.
9 Clara Dong Yu, "Sulla presenza di testi cinesi in Vaticana nel Cinquecento: Nicolas Audebert e il *Vat. estr.-or.* 66," *Miscellanea Bibliothecae Apostolicae Vaticanae* 21 (2015): 559–95.

10 Juan González de Mendoza, *The History of the Great and Mighty Kingdom of China, and the Situation Thereof*, ed. George T. Staunton, trans. Robert Parke, 2 vols. (The Hakluyt Society, 1853), 1: 121; 132; C. R. Boxer, *South China in the Sixteenth Century (1550–1575)* (The Hakluyt Society, 1953), lxxxvii; Brockey, "First China Hands," 79; Tsuen-Hsuin Tsien, *Paper and Printing*, volume 5, part 1 of *Science and Civilisation in China*, ed. Joseph Needham (Cambridge University Press, 1985), 314–17; Lach, *Asia*, 2.2 (1977), 58–59.

11 Lach, *Asia*, 1.2 (1965), 803; 2.1, 13; 39; 2.2 (1977), 47; 53; 2.3 (1977), 515.

12 Lach, *Asia*, 2.2, 64; 358–59.

13 A. F. L. Beeston, "The Earliest Donations of Chinese Books to the Bodleian," *Bodleian Library Record* 4, no. 6 (1953): 304–13; Bert van Selm, "Cornelis Claesz's 1605 Stock Catalogue of Chinese Books," *Quaerendo* 13 (1983): 247–59; Nathalie Monnet, "The Chinese Books in the French Royal Library: An Essential Element in the Birth of Western Sinology," *Études chinoises* 37, no. 2 (2018): 167–87; Françoise Richard, "From China to Cambridge: Modification of Rare Chinese Printed Books within Cambridge College Libraries," paper presented at IFLA WLIC 2017, Wrocław, library.ifla.org/1692/1/088-richard-en.pdf; Arianna Magnani, "The Imported Culture: Who Is the Dummy? Considering 'Agency' in the Circulation of Chinese Books in Europe during the Seventeenth-Eighteenth Centuries," *Annali di Ca' Foscari. Serie orientale* 54, Supplement (2018): 575–94; Thijs Weststeijn, "Just Like Zhou: Chinese Visitors to the Netherlands (1597–1705) and Their Cultural Representation," in *Foreign Devils and Philosophers. Cultural Encounters between the Chinese, the Dutch, and Other Europeans, 1590–1800*, ed. Thijs Weststeijn (Brill, 2020), 104–31; 312–42; David Helliwell, "The Bodleian Library's Chinese Collection in the Seventeenth Century," accessed September 12, 2024, serica.ie/17thcent/bod17th.pdf.

14 Bernardino de Escalante, *Discurso de la navegacion…* (La viuda de Alonso Escribano, 1577), fols. 62–64.

15 Mendoza, *History*, 1: 134–37.

16 José de Acosta, *Natural and Moral History of the Indies*, ed. Jane E. Mangan, trans. Frances M. López-Morillas (Duke University Press, 2002), 329–45; Claude Duret, *Thresor de l'histoire des langues de cest univers* (Société helvetiale caldoresque, 1619).

17 Paolo Maria Terzago, *Musæum Septalianum* (Typis filiorum qd. Elisei Violae, 1664), 128; Ole Worm, *Museum Wormianum* (Ludovic and Daniel Elzevir, 1665), 381–83.

18 Pietro Francesco Scarabelli, *Museo, ò Galeria adunata… dallo studio del Sig. canonico Manfredo Settala…* (Per li Figliuoli del qd. Eliseo Viola, 1666). My references are to the 1677 identically titled expanded edition published by Nicolò and the Viola brothers in Tortona.

19 Daniello Bartoli, *La Cina*, ed. Bice Garavelli Mortara (Bompiani, 1975), 116–19; Scarabelli, *Museo*, 214–17.

20 Bartoli, *Cina*, 119–21; Scarabelli, *Museo*, 217.

21 Carla Tavernari, "Manfredo Settala, collezionista e scienziato milanese del '600," *Annali dell'Istituto e museo di storia della scienza di Firenze* 1, no. 1 (1976): 43–61; "Il Museo Settala 1660–1680," *Critica d'arte* 44, nos. 163–65 (1979): 202–20; "Il museo Settala. Presupposti e storia," *Museologia* 7 (1980): 12–46; Paula Findlen, *Possessing Nature: Museums, Collecting, and Scientific Culture in Early Modern Italy* (University of California Press, 1994), 325–34; Marco Navoni, "L'Ambrosiana e il Museo Settala," in *Storia dell'Ambrosiana. Il Settecento* (Cariplo, 2000), 205–55; Ugo Rozzo, "La biblioteca di Manfredo Settala e la storia delle biblioteche," in *L'organizzazione del sapere. Studi in onore di Alfredo Serrai*, ed. Maria Teresa Biagetti (Sylvestre Bonnard, 2004), 353–81; Marco Rossi, Alessandro Rovetta, eds., *Pinacoteca Ambrosiana*, vol. 6 (Mondadori, 2010), 27–131; Antonio Aimi, "The Exotica of the Settala Museum and Other Northern Italian Collections," in *Turquoise in Mexico and North America: Science, Conservation, Culture and*

Collections, ed. J. C. H. King et al. (Archetype Publications, 2012), 155–64; Antonio Aimi, Vincenzo De Michele, and Alessandro Morandotti, "Towards a History of Collecting in Milan in the Late Renaissance and Baroque Periods," in *The Origins of Museums: The Cabinet of Curiosities in Sixteenth- and Seventeenth-century Europe*, ed. Oliver Impey and Arthur MacGregor (The Ashmolean Museum, 2017), 24–28; Germana Perani, "La *maravigliosa galeria* di Manfredo Settala," *Nuova museologia* 42 (2020): 2–13; "Una biblioteca 'evocata,'" *Nuova museologia* 42 (2020): 28–44.

22 Exotic books are rarely mentioned in such accounts. See John Ray, *Observations Topographical, Moral, & Physiological…* (John Martyn, 1673), 245–46. Ray also writes about a mysterious "China calendar in wood."

23 For Settala's use of the word "*bagatelle*," see, for instance, Biblioteca Ambrosiana, ms. Z387 sup, 18r. For his portrait, see Marco Riccomini, "Daniele Crespi. Ritratto di Manfredo Settala," in *Diafane passioni. Avori barocchi dalle corti europee*, ed. M. Sframeli and E. D. Schmidt (Sillabe, 2013), 108.

24 For the "witty text" formulation, see Findlen, *Possessing Nature*, 34; see also her "Jokes of Nature and Jokes of Knowledge: The Playfulness of Scientific Discourse in Early Modern Europe," *Renaissance Quarterly* 43, no. 2 (1990): 292–331. The "abyss" descriptor appears in Maximilien Misson, *Voyage d'Italie de Monsieur Misson*, 4 vols. (Willem vande Water and Jacob van Poolsum, 1722), 3: 232.

25 Bibliothèque Universitaire de Montpellier, Ms. BU H 267, fols. 68–96; cf. Luca Tosi, "'Raccolse fin da paesi più divisi dal nostro Mondo i tanti miracoli di Natura.' La collezione di Manfredo Settala attraverso le opere e i corrispondenti," in *Milano globale. Il mondo visto da qui* (24 Ore Cultura, 2021), 62.

26 Daniela Bleichmar, "Seeing the World in a Room: Looking at Exotica in Early Modern Collections," in *Collecting Across Cultures. Material Exchanges in the Early Modern Atlantic World*, ed. Daniela Bleichmar and Peter C. Mancall (University of Pennsylvania Press, 2011), 15–30.

27 Antoine Schnapper, *Le Géant, la licorne et la tulipe* (Flammarion, 1988), 11.

28 A cursory investigation suggests the "Indice" reflects the state of the collection sometime between 1651 and 1656.

29 Ms. BU H 267, 70v. These items seem to correspond to objects the catalog lists under numbers 2, 6, 7, and 8.

30 Ms. BU H 267, fols. 68–71.

31 Scarabelli, *Museo*, 375.

32 Germana Perani, "Gli album di disegni del Settala. Alcune considerazioni introduttive," *Nuova museologia* 42 (2020): 17.

33 Silvia Spitta, *Misplaced Objects: Migrating Collections and Recollections in Europe and the Americas* (University of Texas Press, 2009), 37.

34 Biblioteca Estense Universitaria, ms. Campori γ.h.01.21, 11r, 13r, 14r. These may correspond to items listed under numbers 8 and 7 in the printed catalogs. Cf. the 1662 receipt for a house sale held in the Biblioteca Nazionale Braidense (shelfmark 8 10 B 53/5^{1}).

35 Findlen, *Possessing Nature*, 43.

36 R. W. Lightbown, "Oriental Art and the Orient in Late Renaissance and Baroque Italy," *Journal of the Warburg and Courtauld Institutes* 32 (1969): 256–58.

37 Daniela Bleichmar, "The Cabinet and the World: Non-European Objects in Early Modern European Collections," *Journal of the History of Collections* 33, no. 3 (2021): 435–45.

38 Richard Lassels, *The Voyage of Italy…* (Vincent du Moutier, 1670), 127.

39 Arianna Magnani, "*Gewu bu qiu ren*, un'enciclopedia popolare cinese nella biblioteca dei Gesuiti a Genova," PhD diss. (University of Venice, 2019).

40 Benjamin Elman, "Collecting and Classifying: Ming Dynasty Compendia and Encyclopedias (*Leishu*)," *Extrême-Orient Extrême-Occident*, special issue (2007):

131–57; Wu Huey-fang 吳蕙芳, *Wanbao quanshu: Ming Qing shiqi de minjian shenghuo shilu* 萬寶全書:明清時期的民間生活實錄 [Complete Books of Myriad Treasures: Factual Records of Folk Daily Life in the Ming and Qing Periods] (Cheng-ta lishih hsi, 2001); Wang Cheng-hua 王正華, "Shenghuo, zhishi yu wenhua shangpin: wan Ming Fujian ban riyong leishu yu qi shuhuamen" 生活、知識與文化商品: 晚明福建版日用類書與其書畫門 [Life, Knowledge, and Cultural Products: The Late-Ming Fujian-edited Daily-use Encyclopedias and Their Calligraphy and Painting Sections], *Bulletin of the Institute of Modern History, Academia Sinica* 41 (2003): 1–85.

41 For the Ambrosian pages, see Magnani, "*Gewu*," 137.

42 Hsu Huilin 許暉林, "Chaogong de xiangxiang: wan Ming riyong leishu "Zhuyimen" de yuyi lunshu" 朝貢的想像: 晚明日用類書《諸夷門》的異域論述 [Imagining Tribute: The Discourse of the Exotic in the "Barbarians" Section of Late Ming Daily-use Encyclopedias], *Zhongguo wenzhe yanjiu tongxun* 20, no. 2 (2010): 169–92; Yuming He, "The Book and the Barbarian in Ming China and Beyond: The *Luo chong lu*, or 'Record of Naked Creatures,'" *Asia Major*, 3rd ser., 24, no. 1 (2011): 43–85.

43 Magnani, "*Gewu*," 142–43.

44 Id., Magnani, "*Gewu*", 145–47.

45 Tadao Sakai 酒井忠夫, Yoshinobu Sakade 坂出祥伸, and Yūichi Ogawa 小川陽, eds. *Chūgoku nichiyō ruisho shūsei* 中国日用類書集成 [A Collection of Chinese Daily-use Encyclopedias], 14 vols. (Kyūko Shoin, 1999–2004), Zhongguo shehui kexueyuan lishi yanjiusuo wenhuashi 中國社會科學院歷史研究所文化室, ed., *Mingdai tongsu riyong leishu jikan* 明代通俗日用類書集刊 [A Compendium of Popular Daily-use Encyclopedias of the Ming Dynasty], 16 vols (Dongfang chubanshe, 2011).

46 Yan Congjian 嚴從簡, *Shuyu zhouzi lu 殊域周咨錄* [A Record of Sundry Opinions on Remote Realms], ed. Yu Sili 余思黎 (Zhongua shuju, 1993), 716–17.

47 Elman, "Collecting," 152–53; He, "Book and Barbarian," 49.

48 Arianna Magnani, *Enciclopedismo cinese in Europa: percorsi transculturali del sapere tra Seicento e Settecento* (De Ferrari, 2020), 63–73.

49 Ma Jixing 馬繼興, *Zhongyi wenxianxue* 中醫文獻學 [Bibliographic Studies in Chinese Medicine] (Shanghai kexue jishu chubanshe, 1990), 151–54; Cai Zhongzhi 蔡忠志 and Hao Baohua 郝保華, "*Wang Shuhe Maijue* chongtan: yi 'Wuzang' ge wei yanjiu zhongxin" 《王叔和脈訣》重探—以「五臟」歌為研究中心 [Investigating Anew the *Wang Shuhe Maijue*, with a Focus on the "Five Viscera" Song], *Zhongyiyao zazhi* 25, special issue (2014): 213–31.

50 Scarabelli, *Museo*, 123; cf. Terzago, *Musæum*, 129. See also Girard de Rialle, "Une mission chinoise à Venise au XVIIe siècle," *T'oung Pao* 1, no. 2 (1890): 114.

51 Paul Rule, "The Jesuits and the Ming-Qing Transition: How Did Boym and Martini Find Themselves on Opposite Sides?," *Monumenta Serica* 59 (2011): 243–58; Han Qi 韓琦, "Nan Ming shichen Bo Mige de Zhongguo suicong: jiaotu Zheng Andelei shishi kaoshi" 南明使臣蔔彌格的中國隨從——教徒鄭安德肋史事考釋 [The Chinese Attendant of the Southern Ming Envoy Michał Boym. A Historical Investigation of the Christian Andreas Zheng], *Qingshi yanjiu* 1 (2018): 121–26.

52 Harold J. Cook, "Creative Misunderstandings: Chinese Medicine in Seventeenth-Century Europe," in *Cultures in Motion*, ed. Daniel T. Rodgers, Bhavani Raman, and Helmut Reimitz (Princeton University Press, 2014), 215–40; Marta Hanson, Gianna Pomata, "Travels of a Chinese Pulse Treatise. The Latin and French Translations of the *Tuzhu maijue bianzhen* 圖註脈訣辨真 (1650s–1730s)," in *Translation at Work: Chinese Medicine in the First Global Age*, ed. Harold J. Cook (Brill, 2020), 23–57.

53 Li Yun 李雲, *Zhongguo renming cidian* 中醫人名辭典 [A Biographical Dictionary of Chinese Medicine] (Guoji wenhua, 1988), 20; Serge Franzini, "Un texte médical disparu en Chine depuis le XVIIe siècle conservé à la Bibliothèque Nationale de Paris: le 'Maijue zhengyi' 脉訣正義 de Ma Shi 馬蒔," *T'oung Pao*, 2nd ser., 76, no. 1/3 (1990): 49–61.
54 David Helliwell, "More Seventeenth-century Finds," *Serica*, November 28, 2020, serica.blog; id., "Chinese Books in Europe in the Seventeenth-Century," accessed September 12, 2024, serica.ie/17thcent/17theu.html.
55 Pier Francesco Fumagalli, "Sinica federiciana. Il fondo antico dell'Ambrosiana," *Aevum* 78, no. 3 (2004): 744.
56 Cf. Pier Francesco Fumagalli, "Sinica ambrosiana. Il fondo antico: libri, cimeli e documenti," *Aevum* 75, no. 3 (2001): 728.
57 Letter to Antonio Magliabechi, December 23, 1671; Alfonso Mirto, "Manfredo Settala: lettere al fiorentini," *Galilæana* 15 (2018): 242–43. See also letters to Magliabechi, August 22, 1668 (Mirto, "Manfredo Settala," 241–42) and Athanasius Kircher, July 14, 1677 (Archivio della Pontificia Università Gregoriana, ms. 566, 36r).
58 George Grigore, personal communication.
59 For another Boym performance, see Scarabelli, *Museo*, 123.
60 Two beautifully drawn characters (是也) are immediately recognizable. Changing the orientation of the page, several rather twisted white characters spelling part of a final formulation in a prayer as well as what appears to be the name of the person handwriting it, become legible. Reverse-mirroring the resulting image also reveals numerous but barely recognizable characters, including 佛 (the Buddha).
61 Willemijn van Noord, "Between Script and Ornament: Delftware Decorated with Pseudo-Chinese Characters, 1680–1720," *Journal of Design History* 34, no. 1 (2021): 1–20.

Bibliography

Manuscript sources

Archivio della Pontificia Università Gregoriana, ms. 566. (n.d.)
Biblioteca Ambrosiana, ms. Z387.(n.d.)
Biblioteca Estense Universitaria, ms. Campori γ.h.01.21.(n.d.)
Bibliothèque Nationale De France, BnF Chinois 5186. (n.d.)
Bibliothèque Universitaire De Montpellier, Ms. BU H 267. (n.d.)

Primary sources

Acosta, José De. *Natural and Moral History of the Indies*. Edited by Jane E. Mangan. Translated by Frances M. López-Morillas. Duke University Press, 2002.
Barros, João De. (n.d.), *Ásia*. Edited by Hernani Cidade, 4 vols. Divisão de Publicações e Biblioteca. Agência Geral das Colónias, 1945–1946.
Bartoli, Daniello. *La Cina*. Edited by Bice Garavelli Mortara. Bompiani, 1975.
Duret, Claude. *Thresor de l'histoire des langues de cest univers*. Société helvetiale caldoresque, 1619.
Escalante, Bernardino De. *Discurso de la navegacion…* La viuda de Alonso Escribano, 1577.
Giovio, Paolo. *Histoires de Paolo Jovio*. Translated by Denis Sauvage. 2 vols. Guillaume Rouille, 1561.

Kircher, Athanasius. *La Chine d'Athanase Kirchere…* Johannes Janssonius van Waesberge, 1670.
Lassels, Richard. *The Voyage of Italy…* Vincent du Moutier, 1670.
Liu, Ziming 劉子明, ed. *Xinban zengbu tianxia shiyong wenlin miaojin wanbao quanshu* 新板增補天下便用文林妙錦萬寶全書 [Complete Book of Myriad Treasures: Fine Brocade from the Forest of Letters for Use in the World. Newly Carved, Expanded, and Supplemented]. Anzhengtang, 1612.
Mendoza, Juan González De. *The History of the Great and Mighty Kingdom of China, and the Situation Thereof.* Edited by George T. Staunton. Translated by Robert Parke. 2 vols. The Hakluyt Society, 1853.
Mirto, Alfonso. "Manfredo Settala: lettere ai fiorentini." *Galilæana* 15 (2018): 221–65.
Misson, Maximilien. *Voyage d'Italie de Monsieur Misson.* 4 vols. Willem vande Water and Jacob van Poolsum, 1722.
Moscardo, Lodovico. *Note overo memorie del Museo di Lodovico Moscardo…* Paolo Frambotto, 1656.
Ray, John. *Observations Topographical, Moral, & Physiological…* John Martyn, 1673.
Sakai, Tadao 酒井忠夫, Yoshinobu Sakade 坂出祥伸, and Yūichi Ogawa 小川陽, (n.d.), eds. *Chūgoku nichiyō ruisho shūsei* 中国日用類書集成 [A Collection of Chinese Daily-use Encyclopedias]. 14 vols. Kyūko Shoin, 1999–2004.
Sanuto, Marino. *I diarii di Marino Sanuto.* 58 vols., edited by Stefani, Federico et al. Visentini, 1879–1903.
Scarabelli, Pietro Francesco. *Museo, ò Galeria adunata… dallo studio del Sig. canonico Manfredo Settala…* Nicolò e Fratelli Viola, 1677.
Silva, Luís Augusto Rebelo Da (ed.). *Corpo diplomatico portuguez*, vol. 1. Typographia da Academia Real das Sciencias, 1862.
Terzago, Paolo Maria. *Musæum Septalianum.* Typis filiorum qd. Elisei Violae, 1664.
Worm, Ole. *Museum Wormianum.* Ludovic and Daniel Elzevir, 1665.
Yan, Congjian 嚴從簡. *Shuyu zhouzi lu 殊域周咨錄* [A Record of Sundry Opinions on Remote Realms]. Edited by Yu Sili 余思黎. Zhongua shuju, 1993.
Zhongguo, Shehui Kexueyuan Lishi Yanjiusuo Wenhuashi 中國社會科學院歷史研究所文化室, ed. *Mingdai tongsu riyong leishu jikan* 明代通俗日用類書集刊 [A Compendium of Popular Daily-Use Encyclopedias of the Ming Dynasty]. 16 vols. Dongfang chubanshe, 2011.

Literature

Aimi, Antonio, Vincenzo De Michele and Alessandro Morandotti. "Towards a History of Collecting in Milan in the Late Renaissance and Baroque Periods." In *The Origins of Museums: The Cabinet of Curiosities in Sixteenth- and Seventeenth-century Europe*, edited by Oliver Impey and Arthur MacGregor. The Ashmolean Museum, 2017.
Aimi, Antonio. "The Exotica of the Settala Museum and Other Northern Italian Collections." In *Turquoise in Mexico and North America: Science, Conservation, Culture and Collections*, edited by J. C. H. King et al. Archetype Publications, 2012.
Bedini, Silvio A. *The Pope's Elephant.* Penguin Books, 1997.
Beeston, A. F. L. "The Earliest Donations of Chinese Books to the Bodleian." *Bodleian Library Record* 4, no. 6 (1953): 304–13.
Bleichmar, Daniela. "Seeing the World in a Room: Looking at Exotica in Early Modern Collections." In *Collecting Across Cultures. Material Exchanges in the Early Modern Atlantic World*, edited by Daniela Bleichmar and Peter C. Mancall. University of Pennsylvania Press, 2011.

Bleichmar, Daniela. “The Cabinet and the World: Non-European Objects in Early Modern European Collections.” *Journal of the History of Collections* 33, no. 3 (2021): 435–45.

Boxer, C. R. *South China in the Sixteenth Century (1550–1575)*. The Hakluyt Society, 1953.

Brockey, Liam Matthew. “The First China Hands: Iberian Visitors to the Ming Empire and the Forgotten Origins of Sinology.” In *Western Visions of the Far East in a Transpacific Age (1522–1671)*, edited by Christina Lee. Ashgate, 2012.

Cai, Zhongzhi 蔡忠志, and Hao Baohua 郝保華, “*Wang Shuhe Maijue* chongtan: yi ‘Wuzang’ ge wei yanjiu zhongxin”《王叔和脈訣》重探—以「五臟」歌為研究中心 [Investigating Anew the *Wang Shuhe Maijue*, with a Focus on the “Five Viscera” Song]. *Zhongyiyao zazhi* 25, special issue (2014): 213–31.

Cook, Harold J. “Creative Misunderstandings: Chinese Medicine in Seventeenth-Century Europe.” In *Cultures in Motion*, edited by Daniel T. Rodgers, Bhavani Raman, and Helmut Reimitz. Princeton University Press, 2014.

De Ciutiis, Salvatore. *Une ambassade portugaise à Rome au XVIe siècle*. Michele D’Auria, 1899.

Deswarte-Rosa, Sylvie. “Échos de la Chine à Rome en provenance du Portugal au XVIe siècle: Paolo Giovio, Giovanni Ricci da Montepulciano, Frei Bartolomeu dos Mártires.” In *Património cultural chinês em Portugal*, edited by Luís Filipe Barreto and Vítor Serrão. Centro Científico e Cultural de Macau, 2015.

Elman, Benjamin. “Collecting and Classifying: Ming Dynasty Compendia and Encyclopedias (*Leishu*).” *Extrême-Orient Extrême-Occident,* special issue (2007): 131–57.

Findlen, Paula. “Jokes of Nature and Jokes of Knowledge: The Playfulness of Scientific Discourse in Early Modern Europe.” *Renaissance Quarterly* 43, no. 2 (1990): 292–331.

Findlen, Paula. *Possessing Nature: Museums, Collecting, and Scientific Culture in Early Modern Italy*. University of California Press, 1994.

Franzini, Serge. “Un texte médical disparu en Chine depuis le XVIIe siècle conservé à la Bibliothèque Nationale de Paris: le ‘Maijue zhengyi’ 脉訣正義 de Ma Shi 馬蒔.” *T’oung Pao*, 2nd ser., 76, no. 1/3 (1990): 49–61.

Fumagalli, Pier Francesco. “Sinica ambrosiana. Il fondo antico: libri, cimeli e documenti.” *Aevum* 75, no. 3 (2001): 715–36.

Fumagalli, Pier Francesco. “Sinica federiciana. Il fondo antico dell’Ambrosiana.” *Aevum* 78, no. 3 (2004): 725–71.

Han, Qi 韓琦. “Nan Ming shichen Bo Mige de Zhongguo suicong: jiaotu Zheng Andelei shishi kaoshi” 南明使臣蔔彌格的中國隨從——教徒鄭安德肋史事考釋 [The Chinese Attendant of the Southern Ming Envoy Michał Boym. A Historical Investigation of the Christian Andreas Zheng]. *Qingshi yanjiu* 1 (2018): 121–26.

Hanson, Marta, and Gianna Pomata. “Travels of a Chinese Pulse Treatise. The Latin and French Translations of the *Tuzhu maijue bianzhen* 圖註脈訣辨真 (1650s–1730s).” In *Translation at Work: Chinese Medicine in the First Global Age*, edited by Harold J. Cook. Brill, 2020.

He Yuming. “The Book and the Barbarian in Ming China and Beyond: The *Luo chong lu*, or ‘Record of Naked Creatures’.” *Asia Major*, 3rd ser., 24, no. 1 (2011): 43–85.

Helliwell, David. “Chinese Books in Europe in the Seventeenth-Century.” *Serica*, Accessed September 12, 2024. serica.ie/17thcent/17theu.html.

Helliwell, David. "More Seventeenth-century Finds." *Serica*, November 28, 2020. serica.blog.

Helliwell, David. "The Bodleian Library's Chinese Collection in the Seventeenth Century." *Serica*, s.d. Accessed September 12, 2024. serica.ie/17thcent/bod17th.pdf.

Hsu, Huilin 許暉林. "Chaogong de xiangxiang: wan Ming riyong leishu "Zhuyimen" de yuyi lunshu" 朝貢的想像：晚明日用類書《諸夷門》的異域論述 [Imagining Tribute: The Discourse of the Exotic in the "Barbarians" Section of Late Ming Daily-use Encyclopedias]. *Zhongguo wenzhe yanjiu tongxun* 20, no. 2 (2010): 169–92.

Lach, Donald F. *Asia in the Making of Europe*. 3 vols. University of Chicago Press, 1965–1993.

Li, Yun 李雲. *Zhongguo renming cidian* 中醫人名辭典 [A Biographical Dictionary of Chinese Medicine]. *Guoji wenhua*, 1988.

Lightbown, R. W. "Oriental Art and the Orient in Late Renaissance and Baroque Italy." *Journal of the Warburg and Courtauld Institutes* 32 (1969): 228–79.

Ma, Jixing 馬繼興. *Zhongyi wenxianxue* 中醫文獻學 [Bibliographic Studies in Chinese Medicine]. *Shanghai kexue jishu chubanshe*, 1990.

Magnani, Arianna. *Enciclopedismo cinese in Europa: percorsi transculturali del sapere tra Seicento e Settecento*. De Ferrari, 2020.

Magnani, Arianna. "*Gewu bu qiu ren, un'enciclopedia popolare cinese nella biblioteca dei Gesuiti a Genova.*" PhD diss., University of Venice, 2019.

Magnani, Arianna. "The Imported Culture: Who Is the Dummy? Considering 'Agency' in the Circulation of Chinese Books in Europe during the Seventeenth-Eighteenth Centuries." *Annali di Ca' Foscari. Serie orientale* 54, Supplement (2018): 575–94.

Monnet, Nathalie. "The Chinese Books in the French Royal Library: An Essential Element in the Birth of Western Sinology." *Études chinoises* 37, no. 2 (2018): 167–87.

Navoni, Marco. "L'Ambrosiana e il Museo Settala." In *Storia dell'Ambrosiana. Il Settecento*. Cariplo, 2000.

Perani, Germana. "Gli album di disegni del Settala. Alcune considerazioni introduttive." *Nuova museologia* 42 (2020a): 14–23.

Perani, Germana. "La *maravigliosa galeria* di Manfredo Settala." *Nuova museologia* 42 (2020b): 2–13.

Perani, Germana. "Una biblioteca 'evocata'." *Nuova museologia* 42 (2020c): 28–44.

Rialle, Girard De. "Une mission chinoise à Venise au XVIIe siècle", *T'oung Pao* 1, no. 2 (1890): 99–117.

Riccomini, Marco. "Daniele Crespi. Ritratto di Manfredo Settala." In *Diafane passioni. Avori barocchi dalle corti europee*, edited by M. Sframeli and E. D. Schmidt. Sillabe, 2013.

Richard, Françoise. "From China to Cambridge: Modification of Rare Chinese Printed Books within Cambridge College Libraries." Paper presented at IFLA WLIC 2017, Wrocław. library.ifla.org/1692/1/088-richard-en.pdf.

Rossi, Marco, and Alessandro Rovetta, eds. *Pinacoteca Ambrosiana*, vol. 6. Mondadori, 2010.

Rozzo, Ugo. "La biblioteca di Manfredo Settala e la storia delle biblioteche." In *L'organizzazione del sapere. Studi in onore di Alfredo Serrai*, edited by Maria Teresa Biagetti. Sylvestre Bonnard, 2004.

Rule, Paul. "The Jesuits and the Ming-Qing Transition: How Did Boym and Martini Find Themselves on Opposite Sides?" *Monumenta Serica* 59 (2011): 243–58.

Schnapper, Antoine. *Le Géant, la licorne et la tulipe*. Flammarion, 1988.

Spitta, Silvia. *Misplaced Objects: Migrating Collections and Recollections in Europe and the Americas*. University of Texas Press, 2009.

Tavernari, Carla. "Il Museo Settala 1660–1680." *Critica d'arte* 44, nos. 163–65 (1979): 202–20.

Tavernari, Carla. "Il museo Settala. Presupposti e storia." *Museologia* 7 (1980): 12–46.

Tavernari, Carla. "Manfredo Settala, collezionista e scienziato milanese del '600." *Annali dell'Istituto e museo di storia della scienza di Firenze* 1, no. 1 (1976): 43–61.

Tosi, Luca. "'Raccolse fin da paesi più divisi dal nostro Mondo i tanti miracoli di Natura.' La collezione di Manfredo Settala attraverso le opere e i corrispondenti." In *Milano globale. Il mondo visto da qui*. 24 Ore Cultura, 2021.

Tsien, Tsuen-Hsuin. *Paper and Printing*. Volume 5, part 1 of *Science and Civilisation in China*, edited by Joseph Needham. Cambridge University Press, 1985.

Van Noord Willemijn. "Between Script and Ornament: Delftware Decorated with Pseudo-Chinese Characters, 1680–1720." *Journal of Design History* 34, no. 1 (2021): 1–20.

Van Selm, Bert, "Cornelis Claesz's 1605 Stock Catalogue of Chinese Books." *Quaerendo* 13 (1983): 247–59.

Wang, Cheng-hua 王正華. "Shenghuo, zhishi yu wenhua shangpin: wan Ming Fujian ban riyong leishu yu qi shuhuamen" 生活、知識與文化商品: 晚明福建版日用類書與其書畫門 [Life, Knowledge, and Cultural Products: The Late-Ming Fujian-edited Daily-Use Encyclopedias and Their Painting and Calligraphy Sections]. *Bulletin of the Institute of Modern History, Academia Sinica* 41 (2003): 1–85.

Weststeijn, Thijs. "Just Like Zhou: Chinese Visitors to the Netherlands (1597–1705) and Their Cultural Representation." In *Foreign Devils and Philosophers. Cultural Encounters between the Chinese, the Dutch, and Other Europeans, 1590–1800*, edited by Thijs Weststeijn. Brill, 2020.

Wu, Huey-fang 吳蕙芳. *Wanbao quanshu: Ming Qing shiqi de minjian shenghuo shilu* 萬寶全書: 明清時期的民間生活實錄 [Complete Books of Myriad Treasures: Factual Records of Folk Daily Life in the Ming and Qing Periods]. Cheng-ta lishih hsi, 2001.

Yu, Clara Dong. "Sulla presenza di testi cinesi in Vaticana nel Cinquecento: Nicolas Audebert e il Vat. estr.-or. 66," *Miscellanea Bibliothecae Apostolicae Vaticanae* 21 (2015): 559–95.

5

EVLIYA ÇELEBI IN EGYPT

Self-fashioning and the creation of the wondrous

Mahmoud Abdelhamid M. A. Khalifa

In his *Seyahatnâme*,[1] or *Book of travels*, Evliya Çelebi (1611–1682), an Ottoman traveller of the seventeenth century, reflects the world of a Muslim traveller who has wanderlust and is aware of the long tradition of travel writing that reached its peak with the *Travels of Ibn Battuta* (1354). Çelebi begins the tenth volume of his travelogue in the customary way of glorifying God and making the statement of Muslim faith. Çelebi dates his arrival in Egypt as Safar 1380, which corresponds to June 1672. After arriving there, he "looked carefully all over Egypt and was astonished by its marvelous monuments", which could not be found elsewhere. "Who built its monuments and its talismans?" Çelebi asks and states that he had "read many authoritative books and dependable histories".[2] In search of the answer, Çelebi provides a list of the books he had read. The sheer number of classical books in his list betrays long periods of extensive reading, which eventually resulted in a detailed use of the sources in *Seyahatnâme*.[3]

Çelebi was conversant with many academic sources of the period, and it is likely that he used the important bibliographic work entitled *Kaşf az-Zunūn* (dispelling doubt) on Arabic and Islamic books completed by his contemporary Kâtip Çelebi in 1652, twenty years prior to Evliya Çelebi's arrival in Egypt. Mastering Arabic and spending time in Egypt, Çelebi had access to many of the books listed in Kâtip Çelebi's work. He also had with him "seven rare and valuable books of chronicles", given to him at the outset of his journey by a sheikh who wished him well and exclaimed, "go forth, may your affair come out right. A Fatiha!" At the beginning of his travelogue, Çelebi states he had "studied some of those histories" before embarking on a description of his "birthplace, the envy of kings and harbour of vessels, mighty fortress of the

DOI: 10.4324/9781003506928-5

province of Macedon—Istanbul".[4] Çelebi's extensive reading is manifested in his verbatim incorporation of fictional stories, which he copies profusely and uncritically. One example of such indiscriminate reiteration is his claim that crocodiles have no anus.

Travel as self-fashioning

Strictly speaking, most early modern travelogues were not autobiographical. Evliya Çelebi's *Seyahatnâme* can be classified as such a work. Although it deals with personal events, it does not directly address personal issues and development. It has been pointed out that "travelogues that seem to modern eyes very impersonal and un-autobiographical can sometimes serve as a mode of self-fashioning, by which the writer seeks to project to the wider world a desired identity or persona".[5] Travel is a transformative experience in which the self is perturbed, disoriented and transformed by the encounter with the strangeness of the Other. However, Dankoff dismisses such a transformative effect of travel in *Seyahatnâme*. According to him, Çelebi's *Book of travels*, "with its first-person narrative and focus on human geography", is "reminiscent of another tradition of travel writing", which is "the Chinese 'lyric travel account' (*yu-chi*)". Unlike early modern European travel writing, the Chinese accounts focus primarily on domestic travel itineraries and therefore contain little evidence of "the idea of self-fashioning, both of the individual traveller and of the collectivity of home society". Dankoff argues that "this is also the case with the *Book of Travels*" because whenever Çelebi "did venture into the realm of the 'Other' we do not get the impression that any of the encounters was transformative". What is more, Dankoff asserts that Çelebi "usually felt out of his element beyond the Ottoman borders, and seemed most at home among the Tatars, who were partially Ottomanised".[6]

The reason why Dankoff fails to see transformation in Çelebi is that he positions the Ottoman traveller within the paradigm of early modern European self-fashioning, which was oppositional and hostile. According to Stephen Greenblatt, the subjectivity of the natives is being erased when it serves as a site for the European to construct an identity through othering, which denies the identity of the natives through putative marks of otherness, such as cannibalism and human sacrifice. In his book *Renaissance Self-fashioning* (2005), Stephen Greenblatt contends that the same discourse circulated the idea of a centralized and stabilized identity in the face of an "alien" Other. The "Other" served as a vehicle of European identity formation. Greenblatt argues that "self-fashioning is achieved in relation to something perceived as alien, strange or hostile". Furthermore, such "threatening Other—heretic, savage, witch, adulteress, traitor, Anti-Christ—must be discovered or invented in order to be attacked and destroyed".[7] This kind of self-fashioning aimed at reducing and controlling otherness. Çelebi, however, sought to explain and empathize with otherness. He was not threatened by the otherness of Egypt: on the contrary, he was intrigued by its wonders and grotesqueries that enabled him to empathetically critique some of the Ottoman injustices towards Egyptian peasants. Evliya Çelebi's self-fashioning as a metropolitan Muslim traveller and diplomat was rooted not in control but in wonder dissociated from the machinations of power. Far from fashioning himself in opposition to others, he welcomed the multiplicity of the wondrous and the marvellous, an attitude appropriate to a sophisticated Istanbul gentleman and intellectual. He fashioned himself in relation to internal others instead of external ones. In his writing, Çelebi often expresses astonishment, repulsion, fear and disgust. At the same time, like his Arab and Muslim precursor Ibn Battuta, he strove to conquer the strange, wondrous and marvellous through plausible explanations. In this sense, the strange became familiar through his act of mediation, which, while seeking to deliver a plausible explanation, nevertheless had the tendency to produce misleading, inaccurate postulations or outright fantasies. It is evident, however, that his encounters with strangers were identity-forming. As Sara Ahmed convincingly argues, "the term encounter suggests a meeting, but a meeting which involves a surprise and conflict".[8] Ahmed also posits an important question that is relevant to our discussion on Evliya Çelebi's encounters with others: "how does identity itself become instituted through encounters with others that surprise, that shift the boundaries of the familiar, of what we assume that we know"[9] She suggests a radical reevaluation of subjectivity in the context of encounters with the strange by arguing that "given that the subject comes into existence as an entity only through encounters with others, then the subject's existence cannot be separated from the others who are encountered". Furthermore, this means that "the encounter itself is ontologically prior to the question of ontology".[10] Çelebi's self-understanding was unsettled by the numerous experiences he had during his travels, experiences which nevertheless shaped his understanding of himself as much as they shaped his knowledge of the world.

Cemal Kafadar has pointed out that Çelebi "considered himself '*iilem ve ferid-i iidemj*', traveller of the world and unique among men", who nevertheless "worried in his youth whether he would ever be able to free himself of his family and his teachers in order to travel". In the end, as Kafadar puts it, Çelebi left behind a narrative in which we can see, besides "the tracks of his endless voyages" and "the observations of his uniquely perceptive pair of eyes", also the "traces of an intricate web of relations and apertures in the boundaries of his self through which he discovered and defined himself".[11] Çelebi fashioned Egypt as a wondrous space and placed himself at the centre of a narrative in which ordinary events were transformed by the power of his imagination. Nowhere, however, does Çelebi suggest the kind of radical difference between himself and the Other, which was presupposed by his European counterparts. He was, after all, walking on familiar terrain. Discussing an earlier work of geography by Qazwini, Travis Zadeh argues that "wonder talk often flirts ambiguously with problems of being and knowing: the marvel of the real of what could be, or what is all but impossible". In his view, "the particular vocabularies of wonder and meanings that they evoke shed light on formation of ideology and authority, offering insight into the diverse changing configuration of feelings and being in the world".[12] After travels that continued for fifty years, Çelebi was transformed from someone with an elitist Ottoman mentality into a world traveller who traversed cultures and religions.

To give Çelebi and his elitist mentality some historical context, it has been pointed out that the Ottoman Empire "produced a coterie literature centered on the court of Istanbul and closely modeled on classical Arabic examples, whereas the literature produced in the country at large, modeled on Turkish tradition, was never taken seriously by the coterie group and always rejected as 'popular' if refereed to it at all".[13] This is confirmed by scholar Michael Winter who argues that "It stands to reason that the Ottoman 'ulema' were awed by the strength and depth of the Islamic scholarly tradition of Egypt, Al-Azhar in particular".[14] Çelebi encountered both European others and Muslim others, and although he shared the common religious creed with the latter, there were also evident differences. Intrigued by these differences, he presents them varyingly as a matter of fact or as something unfamiliar and grotesque to the extent that they evoke wonder and fear. His elitist identity did not prevent him from admiring and empathising with the Egyptians, and coming from the Ottoman elite, he was capable of critiquing both the Egyptians and the Janissaries. While his project of self-fashioning embraced cultural curiosity about Egyptians, it never resulted in renouncing his Turkish Muslim identity. Çelebi defined himself against the Pharaohs, who were, according to Qur'anic verses, tyrants and briefly discusses the legacy of the Pharaohs, remarking that his contemporary Egyptians were subjected to humiliation and oppression. This observation formed the basis for his sustained critique of the mismanagement, which led to the rise in prices and threw Egyptians into a state of abjection. Çelebi was at once attracted by the greatness of the monuments of the pharaonic civilization and repulsed by the circumstances of the Egyptians of his own era. In general, he defined himself as someone who translated Egyptian culture into Turkish idioms.

The wondrous, as something unfamiliar and grotesque, plays an important role in Çelebi's writing. At times, the wondrous turns into carnivalesque as Çelebi sprinkles his narrative with contrasting passages, which provide comic relief to the overloaded pattern of descriptions of places. Çelebi's critique of the status quo through the use of fictional stories and myths can be better understood in the light of Bakhtin's notions of the grotesque and carnivalesque, which defamiliarize the cultural topos and create a new landscape for the critique of social conditions. In many ways, the wondrous and the grotesque go hand in hand. The disruption of the coherence of the alien helps familiarize it, even if it does not necessarily result in a coherent representation.

In his travelogue, Çelebi translated Egypt for a presumed Turkish elitist audience, resorting to a phantasy, which foreignized the subject instead of domesticating it. In so doing, he gleaned from Arabic sources material, which spiced up the terrifying episodes that constitute his description of Egypt.

Çelebi's matter-of-fact description of an Arab copulating with a crocodile in Egypt tests the reader's credulity. It has been suggested that "earlier compendiums of geography and natural history had long traded in the value of astonishment at the strange and extraordinary, mixing stories that strain credulity with quotidian fare and pedestrian facts".[15] The grotesque and carnivalesque were rhetorical devices employed to respond to readers' expectations for the strange and the macabre, which were often associated with distant lands. At the same time, such additions in the text brought comic relief in the midst of an otherwise factual work of ethnographic quality. Karnes argues that "marvels constitute their own genre, one that crosses disciplinary boundaries and encourages comparisons between fictive and nonfictive texts". She adds that marvels are "unlikely and spectacular possibilities", which "conceptualize a space between truth and falsehood that invites play, experimentation, uncertainty, and, of course, wonder".[16] There is a symbiosis between the factual and the fictional in Evliya Çelebi's *Book of travels*. His detailing of the brothels and night life of Cairo is realistic. According to Karnes, "marvels are things that are unexplained, unexplained because rare, or insufficiently explained because the effect seems mismatched to the cause". Marvels are marked by "a deficiency of explanation" and, she points out, "the idea that wonder results from things not fully understood is very easy to find in medieval sources".[17]

As we will see, the *Book of travels* contains two specific episodes that represent the grotesque and the carnivalesque as defamiliarizing techniques. In the first of these, the episode of human sexual intercourse with a crocodile stands for the grotesque. In the second (carnivalesque) example, Çelebi describes how Egyptian peasants impersonated Janissaries, religious scholars and sheikhs in an episode which undermines military, scholarly and religious power structures.

The wondrous is manufactured as a formal trope of travel narrative and is a formalistic literary allusion. For example, Çelebi's description of his visit to Egypt's oldest mosque (established by the governor Amr Ibn Al-'As) involves a story that relates to the dove theme from the beginnings of Islam. It is purported that when prophet Muhammad emigrated from Madinah to Mecca, he took refuge in a cave outside Mecca, and miraculously, a spider spun a cobweb that closed the opening of the cave and a dove laid eggs at the entrance of the cave.[18] The miraculous actions of the spider and the dove dissuaded the polytheists of Mecca—who were pursuing the prophet—from entering the cave because it appeared uninhabited and untouched due to the cobweb and the eggs at the entrance. This miracle is transformed and used by Çelebi as a trope of wonder when he describes his visit to the "Mountain of birds": an ordinary sight of bats entering a cave and hanging upside down turns into a wonder since it was, according to Çelebi, "a strange quirk of nature that wild creatures and denizens of the air should come of their own volition, latch on to the holes in the rock at the top of this cave, and hang there by their beaks and claws until they die".[19] In Çelebi's narrative, what is ordinary takes on marvellous overtones and is explained as "a wondrous act of divination, an ancient talisman that is effective until today – truly white magic! God be praised that in the course of my travels I have been vouchsafed to witness this tremendous spectacle".[20] He does not fail to remind the reader of the privilege of witnessing marvels which both invite the reader to trust him and flaunt his self as the centre of the narrative.

It is in the tradition of Ibn Battuta that Çelebi is best understood. L. P Harvey argues that "rather than fables"Ibn Battuta's writing consisted of tales which were drawn from "life as it was lived elsewhere, and incredibly far away". This alone "sufficed to enthral his audience". Importantly, however, Harvey points out that "the reasons that Ibn Battuta has been so often seen as out of focus, misunderstood, is that his stories have rarely been taken for what some of them are: not so much direct reflections of life's realities, more distractions from those realities".[21]

Modern notion of fiction is not compatible with fiction in the premodern sense. In Çelebi's time, *Ajaib* (Arabic for wonders) were not necessarily removed from either reality or blatant lies. Sometimes *Ajaib* were important embellishments or decorative tropes that were narrated not in the sense of telling a truth but with the intention of complying with the expectations of audiences who wanted travellers to far-away lands to satisfy their desire for the strange and the wondrous. In this sense, Çelebi was bound by a tradition in which the wondrous, marvellous and strange formed a key part of the genre of travel writing. In the Middle Ages and the early modern period, many elements of the world were enchanted, and what strains our credulity now appeared real to Çelebi's reading audience. Karnes asserts that "when fiction is not categorically cordoned off from the real, there is more room to appreciate objects and phenomena that resist classification, such as marvels", which "blur the boundary between what exists in the mind and what exists in the world". According to her, it follows from this that "invented things might resemble real ones and real things might act like inventions".[22] The narratives of the wondrous and the marvellous were means of world-making. The wondrous, marvellous and strange are located at a distance from the centre, and, for Çelebi, Egypt represented such a place. However, in this case the marvellous was not connected to possession as Greenblatt has argued, since Egypt already was an Ottoman possession. Therefore, in Çelebi's writing, the marvellous, wondrous and strange are disconnected from power and sometimes critical of it. This point is coupled with another strategy used in travel writing: the practice of mitigating the strangeness of the marvellous by placing "easily believable claims alongside those that are harder to believe. Such hybridity had long been a feature of geographies and histories in the West, dating back to Pliny, Strabo, Ctesias, and Herodotus".[23]

Grotesque intimacy: intercourse with animals

As far as animals that live in and around the Nile are concerned, Çelebi narrates ("*Ajaib*") wondrous and strange things. Çelebi relates the unbelievable story about a man copulating with a crocodile in a section that contains observable facts about Nile Crocodiles. The grotesque episode unsettles the borders between the human and the animal world through the creation of a sexual relationship in which the animal seduces humans and demonstrates satisfaction afterwards:

> Our host was a certain Abu Jaddullah, a venerable sheikh, one who had experienced the vicissitudes of fortune, one who belonged to the class of lovers (i.e., those mystically inclined), an upright individual. He related as follows, "When I was young I had a crocodile in the Nile. It was a female. At that time I was making a living by fishing with nets. Sometimes I netted a big fish and set it aside. One day this crocodile swam by. It was a very beautiful specimen. I snipped off some fish heads and threw them to her. She ate them and went on. This continued for several days: I gave her fish like this every day. One day she just strode out of the water, raised her tail and lay on her back. I recalled that our Arabs have sex with crocodiles. So I too girded my loins and went to it. The pleasure drove me out of my mind. Afterward I took a club and propped it between the sand and the crocodile's back, then stood to one side. She wriggled onto her feet, gave me a glance, then went back into the river and cavorted like mad".[24]

Here the grotesquerie is generated by diluting or merging the human and the animalistic. While Çelebi's writing transmits knowledge of the natural world through the authority of a presumed witness, the story turns bizarre and breaks the norms of human civilization. The hybridity that is created by this mixture of human and animal has the effect of creating a disturbing wonder. The threatening crocodile is transformed into a feminine figure, intercourse with which is pleasurable and has medicinal effects. As the story proceeds, the reader learns that the half-girl half-crocodile was a creation of a spell cast on a girl. Çelebi's writing created a world in which beings could pass from the human to the animalistic and back again through powers that remained unknown. In the conclusion of the story, the act of spell was undone by the omnipotent God whose power prevailed magic.

> 'Then one day I happened to go over to one of the islands in the Nile, and my crocodile followed me onto the shore. She walked around a bit, then

> rolled over and expired. Suddenly it grew dark, and when the darkness lifted what should I see: the crocodile's head and torso had transformed into that of a beautiful girl. But her legs and her vulva were still those of a crocodile. It turned out that she was the daughter of the sheikh of the Kunuz Arabs and had been bewitched. By God's command, at the moment she died the spell became ineffective and she resumed her original form. The people of that island and I buried her'. This was the story the sheikh related, and some local residents who were present approved it and bore witness that it was true.[25]

In this passage the borders between human and animal are predicated on supernatural acts of magic. While these acts were transformative, they also resulted in something that represents the grotesque in the *Book of travels*.

The grotesque as social critique

Çelebi uses the grotesque and the carnivalesque as defamiliarization techniques, which transform the familiar into something strange and mind-numbing. Both techniques were used to criticize the existing social order, which, in Çelebi's view, treated the Egyptians unfairly. The carnivalesque is manifest in the description of a wedding party during which the Janissaries—the Turkish ruling elite traditionally associated with masculine qualities—are mockingly feminized by the bride. Both the grotesque and the carnivalesque are present in the description of the torture of a village thief, an episode which criticizes the Ottomans for being oppressive and the peasants for their submissiveness. The grotesque and carnivalesque intertwine in the following passage, which relates to the Ottoman system of selling taxes, a practice that placed peasants under extreme conditions of poverty. The thief asks the *Kashif* (tax-collection officer) to dispatch him to the execution arena in pomp, and the Kashif complies. This indicates a reversal of authority as the peasant who is about to be gruesomely executed for defying Ottoman authority is given agency.

> It is a strange thing that when a thief from the Egyptian peasantry had the bad luck of falling into the hands of a *Kashif* who will kill him, he gives him a few purses and says, 'Take me to the execution area in Cairo in a mocking manner'. No sooner does he request that the *Kashif* accepts the request and takes the purses and hands him to the executioners who take him to the execution arena and crucify him, and skin him from his chest to his back alive while he smokes tobacco and sings *Mawawil* and Khayyam's quartets taking pride in those whom he had killed and crucified. He bears the executioner's breaking of his hands and legs without a single moan. Then they skin him and fill him with straw. They clothe his dirty cadaver and mount it

> on a horse and take him to the Diwan of Egypt shouting jovially that this is the so and so thief. This way, the state collects its debts, and the treasury is filled. All this income is taken from the peasants.[26]

The execution and the lack of burial represent the grotesque. Contrary to Sharia, according to which the body of the deceased must be washed, shrouded and prayed over, the cadaver becomes an object of punitive action. The public humiliation of the mutilated corpse of the thief illustrates the cruelty of a system geared to ensuring the flow of farm tax to the Ottoman Porte. This grotesque portrayal of authority shows how the sanctity of the body (which is enshrined in the Sharia) was violated. The horror is heightened by the grotesque display of the cadaver mounted on a horse. This dehumanization of the body, Çelebi implies, correlated with the dehumanizing economic and social conditions created by lax administration and greedy traders. According to Çelebi, this administrative laxity threw many Egyptians into poverty. There is another example of grotesque punishment that represents extreme dehumanization and breaches the Sharia. Referring to his visit to Tariz in Persia, Çelebi describes "the condemnable customs of the Persians", which he happened to witness:

> They tie together his thumbs and big toes and hang him by a noose on a gibbet, then burn sulphur and donkey's urine until the fumes make the poor man scream. They pry out his shoulder blades and his ribs and collar bone, then crucify him with straps drawn through his back and wind-tapers attached to his shoulders. They crush his testicles with a noose. They shove gorse up his nostrils. They put a burning plate on his head and insert a stylus in his eye. Or they may pull out his penis through his anus, or his anus through his navel so that – God forgive us! – he excretes through his navel.[27]

In the manner of a religious scholar, Çelebi is horrified by the punishments and tortures that violate the Sharia. In another passage he describes a conversation he had with an administrator of torture:

> One day, as the Khan was having these tortures administered in his presence, and was taking pride in them despite our discomfort, I asked, 'My Khan, what is the purpose of torturing people to this degree?' 'My brother', he said, 'these are criminals whose guilt has been established by probationary witnesses and who deserve to die. Instead of putting them to death all at once, we subject them to these torments in order that they may serve as warnings to the heedless. They are rebels against God, like Pharaoh who said I am your supreme Lord (79:24). That is why we torture them'.[28]

Çelebi questioned the practice of torture declaring that it was against Islamic Sharia. In the following passage Çelebi wins the argument without, however, changing the governor's mind:

> 'But these are not canonical punishments according to the Sharia,' I objected. 'Nor are they effective. Punishments laid down by the religious law are bitter; such as those in the definitive Koranic texts, As for the man or woman who is guilty of theft, cut off their hands (5:38), and We decreed for them a life for a life (5:45). If you followed these commandments your subjects would have greater fear of committing those crimes.' 'Truly, you give good counsel,' he replied, acknowledging that I had the better argument'.[29]

Obviously, Evliya is a Sunni Muslim and a scholar who supported the mandatory Sharia law that punishes a thief who reaches a particular value of theft by cutting one of his hands. Yet Çelebi rejects extreme punishments that contravene Sharia law. Commenting on that specific event, Çelebi states that "they practice these kinds of extreme tortures in order to maintain strict law and order, and because their subjects are riotous and heretical. Also, such arbitrary punishments are traditional".[30] In this manner he fashions himself as a courageous companion who risks challenging the judgement of people in authority.

The carnivalesque and the unsettling of the norm.

This public enactment of grotesquerie contrasts starkly with the peasant wedding that mocked and feminized authority in a culture of machismo. The story of the peasant bride who camouflaged alternately as a Janissary, a cavalryman, a scholar and a sheikh is an anecdote that Çelebi probably heard or attended in one of Cairo's numerous coffeehouses. "Humorous shows and farces also took place in the coffeehouses. Johanna de Wild, who travelled to Cairo between 1606 and 1610, talks of burlesque farces commonly played in coffee-houses and of 'comedies' that were acted out in street performances".[31] This is corroborated by Çelebi:

> One or two thousand such men forgather in every coffee house and raise a ruckus beyond description. Day and night a thousand boy dancers prance about with coquettish gestures, catching the hearts of lovers in the traps of their flowing locks. Other groups of entertainers – singers, instrumentalists, ghazal-reciters, itinerant musicians, mimics, eulogists, comedians, mimes and buffoons – are everywhere, conducting as it were sessions of Husayn Bayqara.[32]

Evliya alternates between factual and fictitious styles, between providing exact numbers and using fiction for embellishment. When the peasant bride (a fellahin) disguised as a Janissary, she overturned the established sociopolitical order. This was an act ahead of its time, given the traditional gender roles which assigned a passive role for women. Through laughter the social world of the lowest Egyptian strata is brought closer to the highest echelons of religious and political power. As Bakhtin argues, "laughter demolishes fear and piety before an object, before a world, making of it an object of familiar contact and thus clearing the ground for an absolutely free investigation of it". Furthermore, "laughter is a vital factor in laying down that prerequisite for fearlessness without which it would be impossible to approach the world realistically".[33] Laughter undermines the authority of scholars, religious fatwa and the machoistic Janissaries who were feminized by the peasant bride. The social order was overturned in favour of a bride fighting for her dowery and, at the same time, the carnivalesque nature of the passage in the *Book of travels* disrupts the seriousness of reality while allowing the critique of the prevailing social order. As Bakhtin points out, "laughter has the remarkable power of making an object come up close, of drawing it into a zone of crude contact where one can finger it familiarly on all sides, turn it upside down, inside out, peer at it from above and below, break open its external shell".[34] The Egyptian peasantry is allowed the luxury of poking fun and bringing the oppressive world topsy-turvy:

> Fellahin (rural Egyptians). When they want to marry an Arab girl to a Muslim man, the religious ceremony takes place according to the rule, then on the night of the wedding procession 1,000 or 2,000 relatives and associates and hangers-on assemble in a broad plain. They eat Jews' mallow, okra, taro and cheeses. When this feast is over, and after the evening prayer, a lovely young janissary appears from one side of this plain. With shawl and turban and a motley waistband, a silver-handled knife in one hand and a club in the other, shouting raucously and striking anyone in his path, he makes his way through the crowd, chases away some people and breaks some objects, thrashes with his knife and walks and talks like a drunkard. As this sweetheart approaches, some men get up and stop him, kiss his face and seat him in a corner. Now it turns out that this bandit boy is actually the bride! She stands up again and cries, 'Where is our master?' She is supposed to find her husband and try to kill him.[35]

This episode, which Çelebi represents as a wedding drama of the Fellahin, is best understood in the light of Bakhtin's definition of the carnivalesque: "as opposed to the official feast, one might say that carnival celebrated temporary liberation from the prevailing truth and from the established order; it marked the suspension of all hierarchical rank, privileges, norms, and prohibitions".[36]

> This time the bride is out-fitted as a lovely scholar from a good family, with a red mohair cloak over her shoulders, a scarf round her throat and an

> ulema's turban on her head. As she approaches, some relatives go to greet her and some scholarly types stand up and invite her to sit with them. 'No', she says, 'I have a legal dispute with so-and-so'—meaning her husband.[37]

The show goes on as the protean persona of the pride impersonates more figures from the higher classes:

> Then it is the bride's turn again. This time she is dressed as a cavalryman, with red trousers and a dagger in her belt. She searches for the groom, brandishing the dagger, and again lists all the valuable items in her dowry. The next time she comes dressed as a sheikh and imparts all sorts of sheikhly counsel. When it is the groom's turn he shows up as a dervish, with a staff in one hand and prayer-beads in the other, shaking his head and reciting litanies. In short, during that long night until the glimmer of dawn, bride and groom appear ten times each in various costumes and have all sorts of disputes.[38]

In this carnivalesque staging of various Egyptian classes even sexual mores of society become transgressed. This theme is discussed both realistically (descriptions of brothels) and grotesquely (the crocodile anecdote). In the description of the peasant wedding, sexual mores are carnivalized with implied licentiousness:

> And the bride sits on the knee of any man she likes and they talk and kiss. Meanwhile the plain is adorned with hundreds of torches and thousands of oil lamps; kettledrums and five-layered military bands are played; poems are recited in loud voices, both in classical Arabic and in colloquial; and everyone has a wonderful time. A torch procession comes in from one side, with numerous people on foot accompanying the bride who is mounted on a horse with bejewelled saddle. She is smothered in sable furs and wearing all her gold and finery, a Pharaonic crown with three forked crests on her head, and her face still unveiled. She greets the crowd on either side.[39]

The gap between high and popular culture is bridged as "poems are recited in loud voices, both in classical Arabic and in colloquial".[40] It is not only the social structure that is reversed but also the distinction between oral peasant culture and the scholarly Azharites' culture. The border between both is blurred and "everyone has a wonderful time". The carnivalesque caters for all tastes. Çelebi's travelogue combines two styles of writing, one gleaned from books and the literary culture, the other emulating the popular and oral street culture. This shifting of styles and cultural idioms is manifested in the alternation between the serious, statistical listing of mosques, tax returns and cafés on one hand, and the carnivalesque narration of anecdotes about saints' supernatural deeds and the fantastical description of improbable animal attributes on the other. As the wedding comes to its conclusion, a return to normality is signalled by the re-establishment of the traditional authority. "All the fellahin rise to their feet and return the greeting. Hundreds of boys and girls holding candles and crying 'God save the Sultan!' accompany the bride to the middle of the plain. She rubs her handkerchief on the crown and waves it over the crowd. All the fellahin click their heels".[41]

Grotesquerie of the incomplete

There is, in *Seyahatnâme*, a description of field mice that turns the notion of incomplete into grotesquerie. In the passage, Çelebi refers to a "wonder" in which—"by the will of God"—mice are created from the ground without breeding where "Some of them were extracted before taking their final shape and are half-formed: half a mouse and half dust".[42] While his description of the birth of the field mice defies reason, Çelebi carries on with the anecdote and attacks the sceptical readers accusing them of being *Mutazilites* (a shorthand for rationalists in early Islam). Corroborating his account, Çelebi appeals to the will of God as a reason behind all reasons, but also to the evidence acquired through observation. He writes that he had paid a group of Arabs to get such incomplete field mice, which he then observed closely. After having examined a mouse, Çelebi noticed that part of it was made of half-blood and spirit while the rest was dust and clay. Supporting his case against the potential critique from sceptical readers, Çelebi treats the matter as believable on a different level of truth and argues that the incomplete mice were a manifestation of God's power as evidenced in the Qur'an, which asserts that Adam was created from clay. In the paragraph, Çelebi dismisses the rationalists while celebrating the unlimited power of God. Such Sufi or folklore stories were refuted by the Mutazilites against whom Çelebi constructed his arguments.

The grotesque suspends belief or disbelief by problematizing the borders between the normal and abnormal. It has been argued that the grotesque can be "harnessed as a powerful force to resist the tools of normalization" because "a grotesque figure can disrupt notions of normality in favor of conceptualizing and recognizing broader varieties of being and expression as dignified and respected".[43] This is reflected in Çelebi's constant attempt to pass grotesque scenes as believable by endowing them with respectability. For example, Çelebi relates how he "went into the latrines at the Mosque of Sultan Hasan" and while sitting in there, "some fellow opened the door, stuck in his hand and asked me for alms". Continuing the grotesque story, Çelebi writes that "since I was just wiping myself, I spared nothing but gave him whatever was in my hand. 'May God increase your excrement,' he said and went away".[44] Çelebi's unbecoming behaviour had consequences. After having learned from a friend that the beggar at the mosque was the Pole (the head of the saints of Cairo) who had been testing him, Çelebi writes that "I was stunned. And indeed, I had diarrhea for the next two months. His prayer that God increase my excrement came to pass!" Eventually, Çelebi recovered by giving alms.[45] This grotesque hoax led to consequences as Çelebi sought to validate Sufi *Karamat* or supernatural acts with the elucidation of the cause-and-effect relation between his treatment of the Pole and the consequent diarrhoea. Çelebi used the anecdote to suggest and exemplify a different order of reality open only to Sufi masters, pointing out that any tampering with it could unleash harmful powers.

Çelebi's ways of being and knowing were hybrid. While his descriptions of cities, architecture, trade and logistics are to a large extent secular and pragmatic, he nevertheless adheres to the Sufi supernatural deeds. At the same time, he does not fashion himself as a committed Sufi, as someone who would be ready to relinquish worldly pleasures. Çelebi faced the same dilemma as Ibn Batuta when the latter in vain contemplated the option of forsaking worldly pleasures and travelling in order to become a Sufi ascetic. Dankoff writes that "dervishness" conflicted not only with Çelebi's "acquisitiveness but also with his amiableness and sociability".[46] Like his Arab predecessor, Çelebi was a man who wanted to see the world, not to shun it.

Cannibalism is a further example of the grotesque that is translated into the language of wonder:

> The fact that human flesh is tasty is something I witnessed among the Kalmyk cannibals in the Kipchak Steppe. They eat the corpses of their dead, and they also strangle and eat some of their poor Nogay captives – they do not cut their throats, so as not to lose their blood, but just strangle them, cook them and eat them. The Kalmyks claim that there is nothing tastier than human flesh, snake meat, and pork; and the tastiest part is the pig's tail and the 'tail' or coccyx of humans. There are in fact many men of the Kalmyk persuasion in Turkey who know of this taste. But I myself can testify that human flesh is tasty, for if a man kisses his beloved just once, he finds eternal life and is overjoyed. That is how I know that human flesh is tasty.[47]

Çelebi suspends all sensibility regarding cannibalism and discusses the matter in the manner of a food connoisseur. Being unable to fathom the taste of human flesh, he compares it to the act of kissing one's beloved. Cannibalism is also referred to in the previous passage, which tells the story of three brothers detained by the Portuguese:

> At dinner time, we offered what we had – since the best food is what is readily at hand – consisting of camel meat and chicken. But they refused, saying they had not tasted meat for seven years, and began to weep. When we inquired the reason, they related as follows, 'We were three brothers

> travelling with India-merchant ships from India to Ethiopia. Portuguese infidels seized our ships and took all three of us captive. They imprisoned us in their hold. By God's wisdom, one of our brothers died, and they roasted his body and forced us to eat it. We fed on our brother's flesh for an entire month. At the end of a month, by God's wisdom, as we were sailing in the Indian Ocean, a storm hurled us ashore and our ship broke apart. We fled to a mountain top where we bathed in a stream and performed two prostrations of thanksgiving. And we swore never to eat the flesh of any living thing. That is the reason'.[48]

Cannibalism is interpreted contextually by Evliya. At one end, he tries to rationalize it and even argues for the tastiness of human flesh based on comparison with kissing. The other end is a story that turns cannibalism into a grotesque event whereby a group of Muslim prisoners are forced by the Portuguese to eat human flesh. What is interesting here is that contrary to European self-fashioning, Çelebi does not use both instances to create any civilizational hierarchy whereby those who practice cannibalism as a tradition or as a means of torture are placed on a scale of civilization versus barbarism.

Conclusion

In his *Book of travels*, Evliya Çelebi describes his experiences with a certain ambivalence. He relates grotesque and carnivalesque stories and validates them by referring to the testimony of "eyewitnesses". At the same time, his belief in the omnipotence of God facilitates the assumption and depiction of a world in which everything is possible. With the description of grotesque anecdotes of punishments whose sole aim was to terrify people into submitting to an economic system that degraded them, Çelebi lays down an oblique critique of his Egyptian patrons. His writing reveals a respectful approach towards the wondrous, unfamiliar and grotesque. It is also evident that Çelebi inhabited simultaneously two different worlds, one mundane and the other supernatural. This provided him with a double vision, which portrayed the premodern world from multiple perspectives and with ambivalence and fluidity that bears a resemblance to postmodern pluralism. Çelebi was a great observer of people, not hastening to pass judgement. He subscribed to a notion whereby absolute certain knowledge is not always possible and that makes him explain things through different discourses: Sufi, legal, fantasy and the omnipotence of God, whose power knows no limit. In this sense, his being in the-world or self-fashioning is not definitive. He lived a life seeking the wondrous. However, like all people of all periods, he was limited by the world of discourse in which he lived where the clear limits between observable facts and fantasy were permeable, which turned the common sensical and familiar into a wonder. It was as if he refused to get to the bottom of everything so as not to disenchant his audience and sabotage the wondrous. In this sense, the Other or the unexplainable falls into some in-between space of fact and fiction and loses its threatening unfamiliarity.

Notes

1 In this chapter, I rely on selections from Seyahatnâme translated by Dankoff and Sooyong in, *An Ottoman Traveller: Selections from the book of travels of Evliya Çelebi* (2011). However, for parts that are missing from the above-mentioned edition, I rely on the 2003 Arabic Translation of Volume 10 of Seyahatnâme, quotes from which I translate into English.
2 متولي، أحمد فؤاد and ,سليمان، أحمد السعيد ,عزام، عبد الوهاب ,عوني، محمد علي ,أوليا چلبي، **سياحتنامه**. **مصر**. الإدارة المركزية للمراكز العلمية، مركز تاريخ مصر المعاصر، دار الكتب والوثائق القومية، 2003.
Çelebi, Seyahatnâme, (Arabic Translation of vol. 10), 26.
3 متولي، أحمد فؤاد and ,سليمان، أحمد السعيد ,عزام، عبد الوهاب ,عوني، محمد علي ,أوليا چلبي، **سياحتنامه**. **مصر**. الإدارة المركزية للمراكز العلمية، مركز تاريخ مصر المعاصر، دار الكتب والوثائق القومية، 2003.
Çelebi, Seyahatnâme, (Arabic Translation of vol. 10), 26.
4 Dankoff & Kim 2011, 8.
5 Thompson 2011, 99.
6 Celebi, *An Ottoman Traveller*, XXIII.
7 Greenblatt 2005, 9.
8 Ahmed 2000, 6.
9 Ahmed 2000, 7.
10 Ahmed 2000, 7.
11 Kafadar 1989, 150.
12 Zadeh 2023, 5.
13 Lefevere 2017, 14.
14 Winter 1998, 27.
15 Zadeh 2023, 115.
16 Karnes 2022, 26.
17 Karnes 2022, 12.
18 Dankoff & Kim 2011, xxv, point out that "any cave—or something that just resembles a cave—calls up the Cave of Orphans, which probably refers to the cave near Mecca where the Prophet Muhammad and his companion Abu Bakr took refuge from their enemies".
19 Çelebi, *An Ottoman Traveller*, 418.
20 Çelebi, *An Ottoman Traveller*, 416.
21 Harvey 2007, 10.
22 Karnes 2022, 22.
23 Karnes 2022, 113–114.
24 Çelebi, *An Ottoman Traveller*, 374.
25 Çelebi, *An Ottoman Traveller*.
26 Çelebi, 2003, 434–435.
27 Çelebi, *An Ottoman Traveller*, 64.
28 Çelebi, *An Ottoman Traveller*.
29 Çelebi, *An Ottoman Traveller*, 67.
30 Çelebi, *An Ottoman Traveller*, 67,
31 Hanna 2003, 66.
32 Çelebi, *An Ottoman Traveller*, 406.
33 Bakhtin 2002, 23.
34 Bakhtin 2002, 23.

35 Çelebi, *An Ottoman Traveller*, 406–407.
36 Bakhtin 2004, 686.
37 Çelebi, *An Ottoman Traveller*, 407.
38 Çelebi, *An Ottoman Traveller*, 407–408.
39 Çelebi, *An Ottoman Traveller*, 408.
40 Çelebi, *An Ottoman Traveller*.
41 Çelebi, *An Ottoman Traveller*.
42 Çelebi, 2003, p. 612.
43 Edwards, and Graulund 2013, 21.
44 Çelebi, *An Ottoman Traveller*, 395.
45 Çelebi, *An Ottoman Traveller*, 396.
46 Dankoff 2006, 127.
47 Çelebi, *An Ottoman Traveller*, 368.
48 Çelebi, *An Ottoman Traveller*, 427.

Bibliography

Primary sources

Dankoff, Robert, and Sooyong Kim. *An Ottoman Traveller. Selections from the book of travels of Evliya Çelebi*, Eland Publishing Ltd, 2011.
Evliya, Çelebi, *Seyahatnâme*, 2003 (Arabic Translation of vol. 10).
سياحتنامه مصر. متولي، أحمد فؤاد and أوليا چلبي، عوني، محمد علي, عزام، عبد الوهاب, سليمان، أحمد السعيد.
والوثائق القومية، الإدارة المركزية للمراكز العلمية، مركز تاريخ مصر المعاصر، 2003 دار الكتب.

Literature

Ahmed, Sara. *Strange encounters: Embodied others in post-coloniality*. Psychology Press, 2000.
Bakhtin, Mikhail. “Rabelais and his world”. *Literary theory: An anthology* 2 (2004): 686–92.
Bakhtin, Mikhail. *The dialogic imagination: Four essays*. University of texas Press, 2002.
Dankoff, Robert. *An Ottoman Mentality: The World of Evliya Çelebi* (revised second edition). Brill, 2006.
Edwards, Justin, and Graulund, Rune. *Grotesque*. Routledge, 2013.
Faroqhi, Suraiya, Halil İnalcık, Boğaç Ergene, Fikret Adanır Antonis Anastasopoulos Idris Bostan, Palmira Brummett–Amnon Cohen–Jane Hathaway, Klaus Kreiser–Hans Georg Majer–Ahmet, Yaşar Ocak, and Abdeljelil Temimi. *The Ottoman Empire and its Heritage*. (2007).
Greenblatt, Stephen. *Renaissance self-fashioning: from More to Shakespeare*, University of Chicago Press, 2005.
Hanna, Nelly. *In praise of books: a cultural history of Cairo's middle class, sixteenth to the eighteenth century*. Syracuse University Press, 2003.
Harvey, L. P., *Ibn Battuta: Makers of Islamic Civilization*, New York: IB Tauris 2007.
Kafadar, Cemal. “Self and others: The diary of a dervish in seventeenth century Istanbul and first-person narratives in Ottoman literature”. *Studia Islamica* 69 (1989): 121–50

Karnes, Michelle. *Medieval Marvels and Fictions in the Latin West and Islamic World.* University of Chicago Press, 2022.

Lefevere, André. *Translation, rewriting, and the manipulation of literary fame*. Routledge, 2017.

Thompson, Carl. *Travel writing*. Routledge, 2011.

Winter, Michael. "Ottoman Egypt, 1525–1609". In *The Cambridge History of Egypt*, edited by M. W. Daly, 1–33. Cambridge: Cambridge University Press, 1998.

Zadeh, Travis. *Wonders and rarities: The marvelous book that traveled the world and mapped the cosmos*. Harvard University Press, 2023.

6

EXPLAINING WONDERS

Kashmir in the *Voyages* of François Bernier

Ioana Manea

Based on a nine-year stay in the Mughal Empire, Bernier's *Voyages* enjoyed a wide success amongst contemporary readers and shaped the image of India in the 17th century France.[1] Divided into several parts, the *Voyages* contain, for instance, the account of Bernier's experiences in Kashmir, which he was the first European to visit.[2] Under the spell of Kashmir, the author describes the place by an Indian topos as the "terrestrial Paradise of the Indies".[3] In his article *Un paradis sans miracles: le Cachemire de François Bernier*, Frédéric Tinguely demonstrates that, animated by the libertine intellectual approach to miracles, Bernier aimed to reveal what was hidden behind them and used the unfamiliarity specific to the distant Mughal region in order to open his readers' eyes to see the deception, which was at the core of superstitions.[4] Building on Tinguely's study, my article analyses the explanations that Bernier provides for the wonders that he encountered in Kashmir by taking into account his subscription to the philosophy of Pierre Gassendi, whose disciple he was. The chapter focuses on the epistemological methods that Bernier used in order to reveal the causes of the apparently supernatural phenomena, which he encountered during his stay in Kashmir.

François Bernier: an intellectual figure fond of novelty

François Bernier (1620–1688) was part of a circle of people who contributed to the development of contemporary innovative thought in France. Before becoming himself a defender of early modern science, he was exposed to a series of scientific and social experiences that shaped his intellectual profile. The scholar who exerted the greatest influence over him was Pierre Gassendi (1592–1655), best

DOI: 10.4324/9781003506928-6

known for his eclectic philosophy, which was adopted by many disciples of the scientific revolution and which was based on Epicureanism, scepticism, empiricism and Christianity. Besides arguing against Aristotelianism—which had dominated the European intellectual world since the Middle Ages—Gassendi was also opposed to Cartesianism, which was gaining popularity amongst the contemporary pioneering thinkers.

During the 1640's and 1650's, notwithstanding the interruptions provoked by his travels and studies, Bernier was thrice associated with Gassendi.[5] Between 1641–1648, he acted as Gassendi's secretary while at the same time following his lectures, such as those on astronomy at the Collège royal. After two years of travels through Europe, beginning with 1650, he joined Gassendi again and supported him actively. Between 1651–1653, Bernier wrote two Latin pieces in order to take the side of Gassendi in a polemic that opposed him to Jean-Baptiste Morin, an advocate of geocentrism. Additionally, in 1652 and 1653, together with Gassendi, Bernier performed some astronomical observations, which focused on the moon and the sun's eclipses. He remained faithful to Gassendi until the very end and was present when the latter died in Paris in 1655. Later on, in the last quarter of the 17th century, Bernier contributed to the diffusion of Gassendi's philosophy. This was mainly achieved through the three editions of Bernier's *Abrégé de la philosophie de Gassendi*, published between 1674 and 1684, in versions with gradually altered forms. Despite propagating Gassendi's ideas, the *Abrégé* was not a mere translation and synthesis of the Gassendist *Syntagma philosophicum*, for it was adapted to the intellectual tastes of the worldly contemporary society.[6]

Besides Gassendi, Bernier was acquainted with the intellectual circles, which were formed, for instance, around Mersenne and the Dupuy brothers. While Mersenne's circle took a particular interest in natural philosophy, the group which was animated by the Dupuy brothers was open to a more encyclopaedic type of knowledge.[7] Amongst the intellectuals who were part of the Dupuy circle, Bernier became close to a small group known as the libertines.[8] Amongst them, he especially befriended La Mothe Le Vayer, whose curiosity for distant lands he would try to satisfy by means of a letter included in the *Voyages*, as we shall see. Despite being close to Gassendi, with whom they shared the opposition to Cartesianism, the libertines were more ambiguous than him towards religious dogmas. Heirs to the humanist scholarship tradition, which gave a prominent place to the authors of classical Antiquity, libertines such as La Mothe Le Vayer practised critical thinking, which rejected dogmatism. Despite their unquestionable interest in the evolution of science, they did not contribute directly to it, but limited themselves to questions that concerned especially the epistemological relationship between experience and plausibility. Furthermore, Bernier's studies allowed him to familiarise himself with the methods and the results of the science that took shape during his time. In 1652, he spent a few months in Montpellier, where he obtained a doctoral degree in medicine.

While in Montpellier, he became acquainted with the research of the famous anatomist Jean Pecquet on the blood circulation and the lymphatic vessels.

From his formative years, Bernier maintained a penchant for modern science. Though he himself did not elaborate on any philosophical theories, he resorted to Gassendist epistemology when accounting for his experiences in the Mughal Empire by means of plausibility and practical experiment.

Bernier—a traveller animated by curiosity

Despite his intellectual debt to Gassendi and instead of taking part in the preparation of his master's *Opera omnia* published in 1658, Bernier chose to travel to the Orient. Only a few months after Gassendi's death, Bernier left on a thirteen-year voyage whose end point would be the Mughal Empire. What might at first glance seem disloyalty to Gassendi's memory was in fact a journey motivated by a number of reasons. In addition to Morin's animosity and a possible disagreement with Habert de Montmor, who was responsible for the Gassendist manuscripts, Bernier's Oriental travel may also have been triggered by one of Gassendi's unaccomplished projects.[9] As a matter of fact, together with his friend François Luillier, Gassendi had been planning a journey to the Levant region.

In 1656, Bernier began a journey that would lead him from Egypt to Palestine and eventually to the Mughal Empire. In so doing, unlike other contemporary travellers who authored famous travelogues (for example, Jean-Baptiste Tavernier or Jean Chardin), he did not have any other goals except for satisfying his curiosity. It is worth mentioning that, when Bernier left France, he probably did not consider India as his final destination. It was only in 1658, after having, for safety reasons, to abandon an expedition to Ethiopia—probably designed to search for the sources of the Nile—that he left for India. The Mughal Empire that he reached approximately in 1659 was in a state of turmoil, provoked by a war between Shah Jahan's four sons, who all aspired to their father's throne. After some harsh experiences of the fratricide war, Bernier arrived in Delhi, where he joined the court of Aurangzeb who, as the winner of the conflict, became the new emperor. After having initially fulfilled the position of a doctor, he became a member of the circle around the minister Daneshmend Khan whose learning Bernier appreciated and with whom he engaged in philosophical dialogue. These conversations proved fruitful for both interlocutors, since while they familiarised Bernier with the Indian civilisation, they gave the Mughal official access to the latest European philosophical trends. In addition to his scholarly conversations with Aurangzeb's minister, Bernier also seized the opportunity to travel around India, in the company of the Mughal court and other European travellers, such as the French merchant Tavernier. In so doing, for instance, between 1662 and 1664, he was part of Aurangzeb's entourage during a visit to Kashmir. After several travels and numerous experiences and intellectual conversations, Bernier left India in 1668 and returned to France the following year.

Bernier's *Voyages*—a travel account unlike any other

Known under the generic title *Voyages*, Bernier's travel account is composed of several parts. The first part is dedicated to the Sun King and takes the form of a political narration that deals with some of the main events that took place during the war among the four brothers as well as during the five-year period that followed it. The next part is composed of a series of letters, dedicated either to high-ranking political figures or to erudite friends. Hence, those letters concern various topics, adapted to the addressee's interests: the Mughal politics and economy in the letter to Colbert, the towns of Delhi and Agra in the letter to La Mothe Le Vayer, the Hindu superstitions in the letter to Jean Chapelain, the atoms and the human understanding in the letter to Chapelle, the relation of the journey to Kashmir in the letter to Monsieur de Merveilles. The last part is made of a *mémoire* about the income sources of the Mughal emperor.

In addition to being the result of their author's interests, the diversity of topics that underlie Bernier's *Voyages* can also be explained by the social and intellectual environment that was specific to their creation.[10] Firstly, despite the geographical distance, while being in India, Bernier maintained the connections with some circles in France that were erudite, influential and wealthy. Hence, to a certain extent, his letters were shaped by questions and recommendations made by his epistolary interlocutors.[11] Secondly, after his return to France, he joined Mme de la Sablière's salon, which reunited a worldly society made of men and women from various social backgrounds and spheres of intellectual life. Filled with refined conversation, the salon was a space of literary criticism as it provided authors the opportunity to read their texts aloud to an audience whose feedback then guided the subsequent revision work. Therefore, Bernier's *Voyages* "were born out of an exchange, an epistolary conversation with his correspondents while he was in India, as well as through real conversations with his interlocutors in La Sablière's salon as he compiled these texts upon his return".[12]

Bernier's travel account was published in four volumes between 1670 and 1671 by Claude Barbin, an editor who targeted the worldly society.[13] Bernier's work was not initially known under the title *Voyages*, but under long, elaborate titles that corresponded to each of the four volumes and were shaped according to the tastes and expectations of the worldly society. Arousing a wide interest because of the novelty and wealth of information that it contained, Bernier's book enjoyed a large European circulation, which partly benefited from its translation into other European languages.[14] More precisely, it was translated into English (London, 1671–1672 and 1676), Dutch (Amsterdam, 1672), German (Frankfurt, 1672–1673) and Italian (Milan, 1675). The French edition, which was the most widely read—despite some minor distortions of the original text—was published in 1699 in Amsterdam, in two volumes, under the title *Voyages de François Bernier*.

In writing his work, Bernier did not comply with the classical norms of the travel literature.[15] Therefore, for instance, his *Voyages* are not structured according to the pattern outward journey – stay – inward journey and do not constantly describe the places visited. They do not always avoid ambiguity, since they incorporate a plurality of positions while not always making explicit choices. This results from the fact that they aim at pleasing diverse interlocutors like the libertine La Mothe Le Vayer or influential intellectuals like Chapelain, who nourished Bernier's hopes for a prestigious position after his return to France. Moreover, Bernier abandoned the convention of travel literature to make use of commonplaces about the region that was being described and, instead, writes his account from the perspective of the first-person singular, even when dealing with experiences that he did not witness directly.[16]

Bernier's travel to Kashmir or crossing the Hell

Composed of nine letters dedicated to Monsieur de Merveilles, Bernier's travel account to Kashmir is the only one from the *Voyages* whose title explicitly relates to travel literature: *An account of the voyage made in 1664 with the Great Mughal Aurangzeb and his army: From Delhi, the capital of Hindustan, to Lahore; From Lahore to Bhimbar and from Bhimbar to the realm of Kashmir, which the Mughals usually refer to as paradise on earth*.[17] In addition to dealing at length in the ninth letter with the time spent in Kashmir, the travel account also narrates the journey and the preparations that preceded it. Gradually moving the focus from the court to himself and centring the narration on his personal experience, Bernier brings out the intense suffering that he went through before reaching Kashmir.[18] Drawing on his subjective experience, he describes the desert-like regions that he had to traverse before arriving in Kashmir as an area beyond geographical coordinates and as one that is impossible to live in. The hostility of this region was only increased by the length of time spent in crossing it. Of all the adversities produced by this inhospitable environment, the one created by heat was the gravest, extreme enough to exceed the heat Bernier had experienced in Yemen. In writing about these unwelcoming regions, he elaborates a "poetics of the underworld crossing".[19] For instance, his description of the crossing of the river Chenab recalls the mythological notion of the river Styx as a passage to Hades: "the excellence of its water, with which the main *omrahs* provide themselves, instead of that of the Ganges that they had drunk so far, stops me from believing that it is a river which leads to Hell rather than to Kashmir".[20]

The reference to the underworld is accompanied by a description of unbearable bodily pain caused by the merciless sun: "my entire face, my hands and my feet are peeled and my body is entirely covered by small red pustules that prick me like needles".[21] Indeed, the inhospitable environment transformed each day into a struggle for survival, which marked Bernier's writing with what has been described as "pathetic aesthetics".[22] The tone of his literary style is evident when he writes that "farewell, the ink dries at the end of my feather and the feather falls down my hand. Farewell".[23] That death was a constant risk was demonstrated by the demise of fellow travellers, such as the unfortunate rider who, unable to hide from the sun under a tent, was found lifeless under a tree where he had tried to find some shelter.[24] The menace of death was pervasive to the extent that it eroded Bernier's curiosity towards novel phenomena and their causes, the exploration of which was one of his favourite intellectual activities. When he seeks to explain the extreme heat by the presence of the Kashmir mountains—suggesting that they increased the effect of the sunrays and prevented refreshing wind—he quickly surrenders to pessimism, exhausted and hopeless, and concludes abruptly with a thought about the uselessness of any attempt to account for the ordeal he was going through: "But what is the point of philosophizing and looking for the reasons of what will perhaps kill me tomorrow?"[25]

The journey from Lahore to Bhimpar, the entrance point to Kashmir, gives substance to the narrative, which is marked by a suspenseful characterisation of the protagonist/narrator who gradually turns into a hero. The narrator's hero status is emphasised by his inclusion amongst the happy few who were eventually allowed to enter Kashmir. The journey had the quality of an initiation ritual because not all who took part in it were allowed to its last stage.[26] Concerned about the limited resources of Kashmir, a small kingdom whose exhaustion he wanted to avoid, Aurangzeb permitted only a small number of his fellow travellers to accompany him to the final destination.[27] Meanwhile, the majority of Aurangzeb's entourage had to wait in the mountainous surroundings which bordered the much-desired Kashmir.

Kashmir—the Garden of Eden

In Kashmir, Bernier entered a world drastically different from the one where he almost lost his life. Descending the mountains that bordered his destination, Bernier arrived in a welcoming, "temperate" place, which had a "bearable, cooler and milder air".[28] What is more, despite the geographical distance between Kashmir and Europe, he felt as if he had been taken back to his native land because the vegetation was strikingly similar to that in France.[29]

In contrast to the descriptions of the arid and hostile regions, Kashmir is compared to a garden and depicted with topoi specific to characterisations of Eden. The gardenlike appearance of Kashmir resulted mainly from the abundance of water that flowed from the mountains. The water transformed the mountains into a "*locus amoenus*", which teemed with trees, pastures, livestock, game and bees, while there were hardly any venomous or predatory species harmful to human beings. The mountains resembled those of the "Promised Land" because they were equally "innocent" and "flowing with milk and honey".[30] Furthermore, some of the mountains were extremely tall and seemed to reach farther than the terrestrial world, ostensibly coinciding with the mythological hearth of the gods. Permanently covered by snow and giving the impression of reaching beyond clouds and ordinary fog, their summit was "always tranquil and luminous like that of the Olympus".[31]

Within the protective and nourishing frontiers provided by the mountains, the kingdom of Kashmir was comparable to "some great garden entirely green".[32] Its prosperity resulted from the collaboration between human individuals and nature. For example, the Kashmiris skilfully controlled the water that flowed from the numerous springs and small rivers of the mountains and conducted it to the hills that they inhabited and the rice fields that they cultivated.[33] In so doing, they transformed their country into a fertile land that nourished a multitude of plants. These plants served various purposes and included hemp, saffron, cereals, vegetables, grapes, fruit trees, (water)melons and pot herbs.[34] Bernier based his description of Kashmir as an Edenic Garden on a comparison with Europe, his own world.[35] He analysed his discoveries by means of an analogy, by comparing the new and strange with what he already knew. The association of Kashmir with the Terrestrial Paradise results at least partly from the observation that "everything [in Kashmir] is scattered with our plants and our flowers from Europe".[36] However, the comparison is not entirely favourable to Kashmir. As a matter of fact, Bernier does not attribute the lesser variety of fruit species or their poorer quality in comparison with those in Europe to the land properties, but to the weaker skills of the Kashmirian gardeners. He concludes that what is lacking in Kashmir is the practical knowledge of European gardeners who know how to combine and select plants according to the weather and soil conditions.

Kashmir—Terrestrial Paradise with explainable wonders

At first glance, Bernier's references to wondrous phenomena in Kashmir seem to emphasise the supernatural nature of their causes. For instance, he uses the word "miracle" when describing the lifting of a heavy rock by eleven mullahs and when referring to a spring from which water flowed intermittently.[37] Furetière's *Dictionnaire universel* attributes a religious significance to "miracle", since it establishes a connection between God's omnipotence and a prodigy, which is capable of breaking the natural law.[38] In his dictionary, Furetière brings out the relationship between wonder and divinity, and after defining the former as something extraordinary and impossible to understand, he identifies wonders with "God's works".[39] Against this background, it would be reasonable to ascribe religious or supernatural content to the fact that Bernier (who in vain sought to understand the causes behind the irregularity of two different springs) referred to strangely functioning springs as "wonder / *merveille*".[40] While the miracles and wonders of the kingdom of Kashmir might have implied the biblical paradise, Bernier firmly placed this "terrestrial paradise" in a concrete environment.[41] He locates, for example, the approximate surface of Kashmir with respect to precise geographical coordinates of other Indian towns and regions, as well as the neighbouring countries.

The abundance of water, yet another indicator of Kashmir's flourishing natural life, is emphasised in a founding legend reported by Bernier (knowledgeable of the Persian language, Bernier had access to ancient Hindu chronicles collected by Jahangir, the fourth Mughal emperor). According to the legend, the whole of Kashmir had originally been covered by a lake. It was dried up by a "saint old man, called Kacheb", who opened the way for the water by "miraculously" piercing the Baramula mountain that enclosed the lake.[42] Despite reporting the legend, Bernier is distrustful of its authenticity:

> As far as I am concerned, I would not like to deny that this land has once been covered by waters; it is also said about Thessalia and some other countries. But I find it hard to believe that this opening for the water was the work of a man, because the mountain is very big and very high.[43]

Clearly stating his point of view by speaking in the first-person singular, Bernier argues against the wondrous creation of the land of Kashmir. In so doing, he combines the interplay between the affirmative and negative forms with the nuances inherent in the present conditional and the statement intrinsic to the present indicative. He seems to accept the idea that the land of Kashmir was filled with water, an argument strengthened by the stories about other ancient regions such as Thessalia. However, on the basis of common sense, he rejects the idea of a man strong enough to make a hole in the mountain, which acted as a dam for the water. Therefore, instead of a supernatural explanation, Bernier prefers one which, in favour of common sense, uses comparison, an inference based on reason and imagination as well as on encyclopaedic knowledge pertaining to second-hand testimonies and geophysics. From the two premises according to which, firstly, the region of Kashmir was prone to earthquakes and, secondly, as argued by Arabs living in the neighbouring regions, the strait of Bab-el-Mandeb appeared as a result of the sinking of a mountain during an earthquake, he infers the existence of a mountain that sank into an underground cavern during an earthquake.[44] At the same time, however, Bernier uses a present conditional ("I would rather believe") in order to stress that his conclusion is merely probable because of the uncertainties on which it lies.

During his stay, he had the opportunity to explore several wonders that were attributed to the enchanted land of Kashmir. One of these was the spring of Send-brary, an important pilgrimage place for the Hindus who came from everywhere to "bathe and become sanctified in its miraculous waters".[45] The spring occupied a significant place on the Hindu religious map and the Indians frequently talked about its "miracles".[46] These referred to the fact that, in the month of May, the spring flowed for around three quarters of an hour, three times a day: at the crack of dawn, at noon and at night. Unlike in the case of the founding legend of Kashmir, Bernier chose not to report the "several fables" about the origin of the spring, because he did not find "any shade of truth" in them.[47] Instead of such fiction, he elaborated his own explanation that involved not only concrete elements such as the rotation of the sun and the vegetation covering the mountain, but also the assumption that there was ice accumulating inside the mountain. In order to stress the verisimilitude of his explanation regarding the intermittently flowing spring, he makes use of the present conditional: "Taking all this into account, I have imagined that the heat of the sun, together with the particular position and the inner structure of the mountain, could well be the causes of this alleged miracle".[48] The explanation was based on a scientific process that went through several stages.[49] Firstly, Bernier carefully observed the context of the so-called miracle. The terms that he uses to describe his approach during this phase of the research emphasise his focus on observation: "I carefully observed" and "I noticed".[50] Secondly, he made use of his reason and imagination in order to construct a logical argument based on the information that he had collected during the observation of the position and the configuration of the mountain. Thirdly, he tested his explanation by means of empirical information that went beyond the brief period of time when he could observe the flow of the spring.

Animated by curiosity, Bernier looked into the case of another spring whose wonder consisted in the fact that, allegedly, at the slightest noise produced by a human voice or by a foot hitting the earth, the spring started bubbling and bringing delicate sand to its surface.[51] However, unimpressed by the local stories about it, he elaborated an explanation that started with the empirical observation that the action of the spring was in fact independent from people speaking or walking near it. Building on the topos of modesty—resulting from his failure to find the "real cause"[52] of the phenomenon—Bernier suggests two explanations for the phenomenon. The explanations are both probable and result from an inference based on the joint action of reason and imagination which, in the second case, also involves previous knowledge obtained by the means of analogy. According to the first explanation, the sand accumulates until it is moved (with great effort) by the spring water. Introducing the second explanation with "rather" and thus indicating its greater probability, Bernier infers that like in the case of the artificial fountains, the action of the spring is the consequence of an underground wind.[53]

Of all the miracles investigated by Bernier the one that took place in Baramula had the greatest religious significance. Taking place near the shrine of one of the best-known saint dervishes or *pires*, the "miracle" was deemed "certain" and, according to highly regarded Daneshmend Khan, so powerful that it could convert its narrator to Islam.[54] The wonder was composed of two aspects, the healing of the sick persons and the effortless lifting of a heavy rock on the tips of their fingers by eleven men. Studying the allegedly miraculous healing performed by the shrine, Bernier resorted to the method of observation. In so doing, he noticed that the cause of the miracle was quite ordinary and consisted of great amounts of food, especially meat and rice, available in the kitchen near the mosque. This nourishment attracted like a "magnet" those who "claimed" to be sick.[55] Bernier drew an ironic conclusion that questions the authenticity of phenomena rooted in devotion: "but since I am always unfortunate on similar occasions, he (the *pire*) performed no miracle that day upon the sick".[56]

Bernier also applied the method of observation when investigating the miracle of the heavy rock. He discovered that the eleven mullahs who participated in the exploit were cautious to hide the details of their action behind their *cabayes* or long garments. Alert and encouraged by his first observation, Bernier sharpened his visual sense as demonstrated by his choice of words ("opened my eyes", "watched"): "as for myself, who opened well my eyes and watched it from quite close".[57] In contrast to what the mullahs pretended, and after having realised how great an effort they were facing, Bernier suspected that they also used their thumb and not only the tips of their fingers when lifting the rock. Following the observations that raised his suspicions about the mullahs' honesty, he continued his study of the wonder with an experiment.[58] He bribed a mullah with a few coins and, pretending to be a believer, shouted with the others "*Karamet! Karamet*!" ("Miracle! Miracle!").[59] This gave him the opportunity to replace one of the eleven mullahs and to participate in the performance of the wonder. While lifting the rock, he disturbed the course of the alleged wonder and the rock fell on his side. It was only after he followed the mullahs and used his thumb as well that they managed to lift the rock from the earth. The experience in Baramula enabled Bernier to see the truth about the alleged miracle and led him to an abrupt and malicious conclusion. Fearing for his life, he once again bribed the mullahs but was forced to leave in a hurry and, unimpressed by the apparent holiness of the place, to "abandon there the saint and his miracles".[60]

In some cases, if Bernier failed to provide an explanation for an alleged wonder, he limited himself to reporting the ones that he found most plausible. He does this, for example, when mentioning a hermitage that apparently floated in the middle of the lake Wular, the hunting place of the governor. "According to what is said about it", Bernier writes, this hermitage accommodated a hermit who spent his life inside it while the whole thing floated "miraculously on water".[61] In addition to stressing that he only reported what he had heard, Bernier provides his readers with a selection of stories that were told about the hermitage: "a thousand foolish stories are invented about it, which are not worth being reported, except perhaps for what some told me: that it was one of those ancient kings of Kashmir who, out of curiosity, had it [the hermitage] built on large beams attached to each other".[62] He contemptuously discarded all fictional stories that were likely to highlight the fabulous origins of the hermitage and chose to only mention the explanation that seemed rational and probable and was based on a human invention.

Miracles—sources for "philosophical thoughts"

Bernier's attempts to rationally explain the Kashmirian wonders are a continuation of his "ancient philosophical thoughts".[63] Reminiscent of the past, the expression "philosophical thoughts" refers to the time when Bernier accompanied Gassendi in research on natural phenomena through practical experiments. In line with these past experiments, the reflections that Bernier developed while crossing the mountains that lead to Kashmir can be considered "philosophical thoughts", which, starting from empirical observations, sought to find natural explanations for phenomena that appeared incomprehensible and wondrous. The "philosophical thoughts" on Kashmir were inspired, for example, by the case of two winds that blew from contrary directions and that Bernier observed within a distance of less than 200 paces. More specifically, he writes that while ascending the mountain, he faced a north wind, but when descending, the wind blew from the south. As in his previous musings, Bernier explains the surprising phenomenon with a hypothesis based on probability as suggested, for example, by the subordination conjunction "as if": "as if this mountain had pushed from all sides vapor from its depths which, when getting out, produced a wind that descended and took its course in these two opposed valleys".[64]

Besides the example of two opposing winds, Bernier's "philosophical thoughts" explore the case of an old hermit who "performs miracles".[65] Residing on the top of the Pir Panjal mountain, the hermit could be associated with a guardian figure or a demiurge.[66] Having allegedly chosen his place of residence in the time of Jahangir, the fourth Mughal emperor, the hermit was ascribed with miraculous abilities, which consisted of provoking thunders or storms of hail, snow, rain and wind. After bribing the hermit, Bernier learns from him that the real cause of the miraculous storms was the combination of noise and the peculiar qualities of the place.[67] The hermit had assumed the task of trying to stop passers-by from making noises, which could have dramatic, naturally caused effects on the mountain top (as the great Jahangir himself had experienced). The hermit's story is corroborated by a comparison with another phenomenon: the heavy rain that disturbed the beauty and serenity of the lake Sang-safed, fully covered by flowers in summertime.[68] This dramatic event was produced by the stirring of the air by the noise from the people assembled around the lake. This was the experience, for example, of the emperor Shah Jahan, who almost lost his life because of the rain. When reflecting on these two phenomena—which at first sight seemed supernatural—Bernier contented himself with hearsay that offered a sufficiently plausible explanation.

Further connecting his experiences of Kashmirian wonders with the rational quest specific to philosophy, Bernier used some of the wonders as examples in his *Abrégé de la philosophie de Gassendi*. It should be mentioned that instead of simply limiting himself to translating his master's work, Bernier revised the text to make it less abstract and more closely related to the concrete reality.[69] To this end, he adjusted some of Gassendi's concepts and clarified them with examples drawn from his *Voyages*. For instance, Bernier adds a brief account of the Kashmirian spring, which flowed three times a day, in order to complement Gassendi's discussion on the spring in Colmars (Provence), which flowed intermittently and approximately eight times an hour.[70] While Bernier describes these springs as "admirable" and "marvelous", his deliberation is void of religious and supernatural connotations and focuses on the search for natural causes.[71] His discussion combines interrogation, negation and the present conditional, the use of which implies that, due to the difficulty of the task, the causes belong to the realm of verisimilitude: "or to guess what is the cause of this interruption to the flow, is certainly not too easy. Yet could we not say that…".[72] Bernier's investigation started with a visual observation and concluded with the rational explanation according to which the drops that accumulated on a ceiling gradually fell down through an underground hole where they eventually overflowed and, through a channel beneath, fed the sporadic spring.

The discussion on the old hermit in the "philosophical thoughts" and the references to the Kashmirian spring in the *Abrégé* demonstrate that Bernier sought to find rational—instead of supernatural—explanations for wonders. Furthermore, Bernier's references to the Kashmirian wonders do not reiterate the commonplaces of the "tradition of the *mirabilia*" in travel literature.[73] While not explicitly referring to Gassendi in his discussion on the miracles, Bernier was in all likelihood influenced by his teacher's eclectic philosophy and its rationalistic tendencies drawn from scepticism, Epicureanism and empiricism.[74] This is reflected in Bernier's application of empirical experience provided by the senses, especially the faculty of sight. Besides visual observation, Bernier corroborated his conclusions with comparisons between similar experiences, which occurred at different times and places and by gathering as much information as possible about different aspects of the same phenomenon. For Bernier, the information provided by the senses was the foundation of reasoning and of the utilisation of both the intellect and the imagination. The conclusions that he reached on the basis of sense perception and rational processes belonged to the realm of probability. More concerned with what is convincing than with the truth, probability is never set in stone. Therefore, a probable explanation resembles a hypothesis and can coincide with another probable postulation.

Conclusion

Although remaining within the realm of probability, Bernier's explanations of the Kashmirian wonders were rooted in nature and excluded any involvement of the supernatural. For him, the incontestable charm of Kashmir was based on it being a large garden, similar to those in Europe. He accounts for the paradisiacal characteristics of Kashmir by the relative skills of its people and the generosity of its nature, which supplies a mild climate and an abundance of water. By exercising his critical thought and applying an epistemological method developed under the influence of Gassendi, Bernier grounded his explanations of Kashmirian wonders in the combination of three elements: empirical observation, probabilism and the cooperation between reason and imagination.

Notes

1 Beasley 2018.
2 For Bernier having been the first European to visit Kashmir (as mentioned in a letter by Monsieur de Merveilles), see Requemora-Gros 2012, 812.
3 Bernier, *Les voyages*, 410: "*le Paradis terrestre des Indes*". Here and afterwards, all the translations from French into English are mine.
4 Tinguely 2006, 66–67.
5 For the biography of François Bernier, see Moreau-Vittard 2015, 247–248; Tinguely 2008, 9–11; Pintard 2000, 384–387.
6 Moreau-Vittard 2015, 248–249.
7 Hamou 2015, 61–66.
8 Moreau-Vittard 2015, 49–54.
9 Tinguely 2008, 11–17; Murr 1991, 239–311.
10 Beasley 2018, 29–71; Tinguely 2008, 18–20.
11 Tinguely 2008, 19.
12 Beasley 2018, 46.
13 Beasley 2018, 40–47.
14 Tinguely 2008, 22.
15 Requemora-Gros 2012, 816
16 Murr 1991.
17 *Relation du voyage fait en 1664 à la suite du Grand Mogol Aurangzeb, allant avec son armée: De Delhi, capitale de l'Hindoustan, à Lahore; de Lahore à Bhimbar et de Bhimbar au royaume de Cachemire que les Mogols appellent ordinairement le Paradis terrestre des Indes*. For the analysis of the title as well as its translation into English, see Beasley 2008, 45–46.
18 Bedel 2021, 120–124.
19 Bedel 2021, 120: "*poétique de la traversée de l'Enfer*".
20 Bernier, *Les voyages*, 398: "*l'excellence de son eau, dont les grands omerahs se chargent au lieu de celle du Gange dont ils ont bu jusqu'à présent, m'empêche de croire que ce ne soit quelque fleuve pour passer aux Enfers plutôt que pour passer à Cachemire*".
21 Bernier, *Les voyages*, 400: "*tout mon visage, mes mains et mes pieds sont pelés, et mon corps est tout couvert de petites pustules rouges qui me piquent comme des aiguilles*".
22 Bedel 2021, 123: "*esthétique pathétique*".
23 Bernier, *Les voyages*, 401: "*Adieu, l'encre se sèche au bout de ma plume et la plume me tombe de la main. Adieu*".

24 Bernier, *Les voyages*, 400.
25 Bernier, *Les voyages*, 398: "*Mais à quoi me sert-il de philosopher et de chercher les raisons de ce qui me tuera peut-être demain?*"; Tinguely 2006, 59.
26 Bedel 2021, 123–124.
27 Bernier, *Les voyages*, 402.
28 Bernier, *Les voyages*, 414: "*ce m'a été une chose assez surprenante de me voir… passer d'une zone torride à une tempérée, car nous n'eûmes pas plutôt monté cette affreuse muraille du monde… qu'en descendant de l'autre côté nous trouvâmes un air supportable, plus frais, plus doux et plus tempéré*".
29 Bernier, *Les voyages*, 414.
30 Tinguely 2006, 60–61. Bernier, *Les voyages*, 406: "*on peut dire que ce sont des montagnes innocentes et découlantes de lait et de miel, comme étaient celle de la Terre de promission*".
31 Bernier, *Les voyages*, 406: "*dont le sommet en tout temps demeure couvert de neige, et qui paraît au-dessus des nuages et des brouillards ordinaires, toujours tranquille et lumineux aussi bien que l'Olympe*".
32 Bernier, *Les voyages*, 407: "*quelque grand jardin tout vert*".
33 Bernier, *Les voyages*, 406.
34 Bedel 2021, 111 and Bernier, *Les voyages*, 407.
35 Bedel 2021, 112.
36 Bernier, *Les voyages*, 407: "*Tout y est parsemé de nos plantes et de nos fleurs d'Europe*".
37 Bernier, *Les voyages*, 421: "*Daneshmend-Khan, assez content de mon voyage, m'en fit entreprendre un autre pour aller voir un miracle assuré*"; Bernier, *Les voyages*, p. 419: "*je me suis imaginé que la chaleur du soleil, avec la situation particulière et la disposition intérieure de la montagne, pourraient bien être les causes de ce prétendu miracle*".
38 Furetière 1701, 2nd vol., first article "*miracle*": "*Ouvrage de la toute-puissance de Dieu… Miracle, signifie quelquefois un prodige qui nous surprend par sa nouveauté; mais dans un sens plus exact, et philosophique, il signifie, un effet qui n'est point une suite des lois naturelles*". Here and afterwards, I have modernised the orthograph, but kept the original punctuation.
39 Furetière, 1701, 2nd vol., first article "*merveille*": "*Chose rare, extraordinaire, surprenante, qu'on ne peut guère comprendre. Toutes les œuvres de Dieu, sont des merveilles inconcevables*".
40 Bernier, *Les voyages*, 419: "*je fis tous mes efforts pour trouver les raisons de cette merveille*"; 424: "*Or la merveille, dit-on, consiste en ce que le moindre bruit qu'on fasse en parlant, ou en frappant du pied contre terre, meut l'eau et la fait couler et bouillonner comme j'ai dit*".
41 Bernier, *Les voyages*, 410: "*le Paradis terrestre des Indes*".
42 Bernier, *Les voyages*, 407: "*ce fût un certain pire, ou saint vieillard, nommé Kacheb, qui donna issue aux eaux en coupant miraculeusement la montagne de Baramula*".
43 Bernier, *Les voyages*, 407: "*Pour moi, je ne voudrais pas nier que toute cette terre n'eût autrefois été couverte d'eaux ; on le dit bien de la Thessalie et de quelques autres pays. Mais j'ai de la peine à croire que cette ouverture soit l'ouvrage d'un homme, parce que la montagne est très large et très haute*".
44 Bernier, *Les voyages*, 407: "*je croirais plutôt*".
45 Bernier, *Les voyages*, 418: "*ils viennent là de toutes parts en pèlerinage pour se baigner et sanctifier dans cette eau miraculeuse*".
46 Bernier, *Les voyages*, 418: "*les merveilles (car c'est ainsi qu'on parle) d'une fontaine qui est de ce côté-là*".
47 Bernier, *Les voyages*, 418–419: "*Ils font sur l'origine de cette eau plusieurs fables que je ne vous rapporterai point, parce que je n'y vois aucune ombre de vérité*".
48 Bernier, *Les voyages*, 419: "*Tout cela considéré, je me suis imaginé que la chaleur du soleil, avec la situation particulière et la disposition intérieure de la montagne, pourraient bien être les causes de ce prétendu miracle*".

49 For a reading that emphasises the dependence of Bernier's explanation on the "experience of senses, reason and imagination", see Tinguely 2006, 63.
50 Bernier, *Les voyages*, 419: "*Je considérai attentivement… je remarquai*".
51 Bernier, *Les voyages*, 423–424.
52 Bernier, *Les voyages*, 424: "*Pour vous en dire à cette heure la véritable cause, il faudrait y avoir mieux pensé que je n'ai fait*".
53 Bernier, *Les voyages*, 424.
54 Bernier, *Les voyages*, 421.
55 Bernier, *Les voyages*, 422.
56 Bernier, *Les voyages*, 422 : "*Mais comme je suis toujours malheureux en semblables occasions, il [le pire] ne fit point de miracle ce jour-là sur les malades*".
57 Bernier, *Les voyages*, 422: "*Pour moi qui ouvrais bien les yeux et qui y regardais de bien près*".
58 Bedel 2021, 114.
59 Bernier, *Les voyages*, 422 ; Tinguely 2006, 65.
60 Bernier, *Les voyages*, 423: "*(je) laissai là le saint et ses miracles*".
61 Bernier, *Les voyages*, 423: "*Au milieu de ce lac, il y a un ermitage avec son petit jardin qui, à ce qu'on dit, flotte miraculeusement sur l'eau et où l'ermite passe sa vie sans en sortir*".
62 Bernier, *Les voyages*, 423: "*L'on fait encore mille sots contes là-dessus qui ne méritent pas d'être rapportés, hormis peut-être ce que quelques-uns m'ont dit : que ce fut un de ces anciens rois de Cachemire qui par curiosité le fit bâtir sur de grosses poutres attachées les unes aux autres*".
63 Bernier, *Les voyages*, 416–417 : "*anciennes pensées philosophiques*".
64 Bernier, *Les voyages*, 417: "*comme si cette montagne eût poussé de tous côtés une exhalaison de ses entrailles qui, venant à sortir, eût fait un vent qui eût descendu et pris son cours dans ces deux vallons opposés*".
65 Bernier, *Les voyages*, 417: "*il fit des miracles*".
66 Bernier, *Les voyages*, 417.
67 Bernier, *Les voyages*, 417–418.
68 Bernier, *Les voyages*, 424.
69 Moreau 2013, 89–102.
70 Bernier, *Abrégé de la philosophie de Gassendi*, 139–140. It is worth mentioning that in the 1684 edition he eliminates the example of the two contrary winds, which he had inserted in a chapter about winds in the previous edition (Bernier, *Abrégé de la philosophie de Gassendi*, Anisson and Posuel, 1678, 166).
71 Bernier, *Abrégé de la philosophie de Gassendi*, 138: "*Ce qui paraît encore plus admirable, c'est cette autre fontaine qui se trouve dans les montagnes de Provence, et qu'on appelle la fontaine de Colmars*"; 139: "*le même se pourra peut-être dire à proportion de cette autre merveilleuse fontaine des Pyrénées qu'on appelle Lers*".
72 Bernier 1684, 138: "*Or de deviner quelle est la cause de cette interruption de flux, c'est assurément ce qui n'est pas trop aisé. Cependant ne pourrait-on pas dire que*".
73 Moreau 2013.
74 Paganini 2023, 75–105; Bellis 2017, 125–152.

Bibliography

Primary sources

Bernier, F. *Abrégé de la philosophie de Gassendi*. 5th vol. chez Anisson, Posuel and Rigaud, 1684 (2nd ed.).

Bernier, F. *Un libertin dans l'Inde Moghole. Les voyages de François Bernier (1656–1669)*, edited by Frédéric Tinguely, Adrien Paschoud and Charles-Antoine Chamay. Chandeigne, 2008.

Furetière, Antoine. *Dictionnaire universel*. 2nd vol. A. et R. Leers, 1701 (2nd ed.).

Literature

Beasley, Faith. *Versailles Meets the Taj Mahal. François Bernier, Marguerite de La Sablière, and Enlightening Conversations in Seventeenth-Century France*. University of Toronto Press, 2018.

Bedel, Mathilde. *Mirabilia Indiae. Voyageurs français et représentations de l'Inde au XVIIe siècle*. Classiques Garnier, 2021.

Bellis, Delphine. "Nos in Diem Vivimus: Gassendi's Probabilism and Academic Philosophy from Day to Day". In *Academic Scepticism in the Development of Early Modern Philosophy*, edited by Plinio Junqueira Smith and Sébastien Charles. Springer, 2017.

Hamou, Philippe. "Les Sciences". *In Dictionnaire des philosophes français du XVIIe siècle. Acteurs et réseaux du savoir*. Edited by Luc Foisneau with the collaboration of Élisabeth Dutartre-Michaut and Christian Bachelier. Translated by Delphine Bellis, Luc Foisneau and Claire Gallien. Classiques Garnier, 2015.

Moreau, Isabelle. "François Bernier: Philosophers' Fictions / Traveller's Visions". In *Method and Variation. Narrative in Early Modern French Thought*, edited by Emma Gilby. Routledge, 2013.

Moreau-Vittard, Isabelle, "'Libertins' et esprits forts" and "Bernier, François". In *Dictionnaire des philosophes français du XVIIe siècle. Acteurs et réseaux du savoir*. Edited by Luc Foisneau with the collaboration of Élisabeth Dutartre-Michaut and Christian Bachelier. Translated by Delphine Bellis, Luc Foisneau and Claire Gallien. Classiques Garnier, 2015.

Murr, Sylvia. "Le politique 'au Mogol' selon Bernier: appareil conceptuel, rhétorique stratégique, philosophie morale". In *De la royauté à l'État dans le monde indien*, edited by Jacques Pouchepadass and Henri Stern. Éditions de l'École des hautes études en sciences sociales, 1991.

Paganini, Gianni. "Gassendi's Interplay between Skepticism and Empiricism". In *Pierre Gassendi. Humanism, Science, and the Birth of Modern Philosophy*, edited by Delphine Bellis, Daniel Garber and Carla Rita Palmerino. Routledge, 2023.

Pintard, René. *Le Libertinage érudit dans la première moitié du XVIIe siècle*. Slatkine, 2000 (3rd ed.).

Requemora-Gros, Sylvie. *Voguer vers la modernité. Le voyage à travers les genres au XVIIe siècle*. Presses de l'université Paris-Sorbonne, 2012.

Tinguely, Frédéric. "Un paradis sans miracles: le Cachemire de François Bernier", *Études de lettres*, no 3 (2006): 55–69.

Tinguely, Frédéric "Introduction". In *Un libertin dans l'Inde Moghole. Les voyages de François Bernier(1656-1669)*, edited by Frédéric Tinguely, Adrien Paschoud and Charles-Antoine Chamay. Chandeigne, 2008.

7

FROM STRANGE TO FAMILIAR

Ottoman Eyes in Paris in the Eighteenth Century

Ipek Bozkaya

> *Every window was crowded with men, women, and children. Since they had never seen Ottomans before, they were eager to see what kind of people they were.*[1]

As an instigator of curiosity, wonder has always played an important role in the history of philosophy. Its significance is equally noteworthy in the history of emotions and personal sentiments of individuals living as subjects of past empires and states. On a spring day in Paris in 1721, it was curiosity that made Parisians crowd the windows of their houses so that they could witness a group of Ottomans pass through the streets. This was the arrival of the Ottoman delegation of nearly a hundred people led by the Ottoman envoy Mehmet Çelebi (c. 1670–1732). Long before emerging as one of the world's most frequented destinations, Paris hosted Çelebi's transformative expedition, which indicated the Ottoman Empire's endeavour to gain an understanding of Western culture and fostered mutual curiosity between the East and the West. The envoy's arrival in Paris evoked equal amazement in the Parisians and the Ottoman visitors. Mehmet Çelebi spent months visiting the opera, the textile workhouse, mirror workshops, houses, gardens, palaces and the architectural monuments while the French observed the foreign dignitary with keen interest. While scholars have studied various aspects relating to Mehmet Çelebi and his *Sefaretname*, this chapter focuses on curiosity, wonder and astonishment aroused by his arrival in Paris.[2]

DOI: 10.4324/9781003506928-7

Western Curiosity and Ottoman Encounters

Wonder and curiosity played an important role in early modern travel and are useful concepts for scholars who explore the rich interconnections between past individuals and societies.[3] Given this significance, the history of travel in the context of diplomatic engagement and cultural exchange goes far beyond the early modern period. While there were many who, like Marco Polo in the late thirteenth century, set out to travel from West to East out of curiosity and a desire to understand different cultures, curiosity-driven journeys from East to West were fewer and started later. The numerous Western visitors—ambassadors, travellers, merchants and missionaries—who explored the Ottoman region kept detailed records and published them in Europe, creating a body of literature that increased knowledge about the Ottoman Empire in the West. Some of these travellers were motivated by curiosity to learn and acquire new knowledge. For example, Sir Henry Blount states at the beginning of his travelogue *A Voyage into the Levant* (1634) that "intellectual complexions have no desire so strong, as that of knowledge".[4] In his travelogue, which was subsequently translated into French, Blount instructs travellers by outlining the necessary mental and psychological processes, the methodology that enables accurate criticism, and the correct way to merge curiosity and interpretation. Blount travelled east because his previous trips to Spain, France and Italy had not taught him anything new ("since these were Christian countries"). He wanted to discover the unknown and challenge the stereotypical views about the Turks. In his words, his plan was:

> to observe the religion, manners and the policie of the Turks, not perfectly, (which were a task for an inhabitant rather than a passenger) but to go further, as might satisfy this scruple, (to wit) whether to an impartial conceit, the Turkish way appears absolutely barbarous, as we are given to understand, or rather an other kind of civility, different from ours, but no less pretending.[5]

During his journey, Blount observed the Turks' way of life and the history and the structure of the "Turkish" empire.[6]

Also, European captives detained in the Ottoman Empire turned their experiences into travel accounts after their captivity had ended, and these accounts were among the most popular publications of the time. For instance, in his work published in 1544, Bertalan Gyurgyevics gives an eyewitness description of the conditions of slavery experienced by Christians in Ottoman lands, besides discussing Ottoman traditions, morals, religious ceremonies and matters pertaining to everyday life. His book was published in several editions and translated into French and other European languages.[7] Johannes Schiltberger, a Bavarian traveller who left his country in 1394, described in his *Reisebuch*

(Travel Book) his experiences as a prisoner of war (1396–1427) and various things he had observed in Muslim countries. One of the most important geographical descriptions of the Muslim region, Schiltberger's *Reisebuch* was published multiple times after Johannes Gutenberg opened the first printing press in 1450.[8] In addition, Georgius of Hungary, a priest who was captured by the Turks in 1438, wrote *Tractatum de Moribus, Conditionibus et Nequitia Turcorum* (1481), an important early description of Turkish culture and religion. More broadly, in European discourse, the East has often been portrayed as a mystical Other, a figure from a fantastic, exotic, and dreamlike realm.[9] On other occasions, the Ottomans or the Turks were seen as a threat.

Whether they were captives, leisure travellers, or on diplomatic missions, Europeans transformed their experiences in the Ottoman lands into travel writing, which provided detailed information about the Turks and their culture. The Ottomans, in contrast, seem to have lacked such curiosity as they did not produce written observations about the places they visited. This apparent lack of interest was at least partly due to the fact that the Ottomans considered themselves superior to other European nations. The seventeenth-century Ottoman explorer Evliya Çelebi (1611–85) was an exception to this: spending most of his life travelling, his writing reveals explicit curiosity towards new, unfamiliar places. At the beginning of his travelogue *Târih-i Seyyâh Evliyâ Efendi*, Çelebi humorously recounts how a dream in which he met the prophet Muhammad inspired his desire to travel. While most Muslims would have uttered "intercession!" (şefâ'at) in front of the prophet, Çelebi exclaimed "travelling!" (seyâhat).[10] In terms of Ottoman travel writing, Evliya Çelebi's travel book and Mehmet Çelebi's 1721 embassy account stand out as perceptive, literary descriptions of foreign cultures. Unlike Evliya Çelebi, who travelled to satisfy his personal curiosity, Mehmet Çelebi's journey to France was instigated by the needs of the Empire.

French Wonder at the Ottoman Delegation

The political decline of the Ottoman Empire made its leaders and representatives interested in the West.[11] The prestige that the empire had lost in the treaties of Carlowitz (1699) and Passarowitz (1718) prompted the Ottomans to perceive Europe as their equal and as something worthy of their attention.[12] For many centuries, the Europeans explored the East and established embassies in the Ottoman territories, while the Ottoman Empire refused to study Europe.[13] As the empire used unilateral diplomacy to maintain the superiority it had gained through war and refused to accept any Christian European state as an interlocutor, it remained, to a large extent, unaware of the developments in the West.[14] However, from the eighteenth century onwards, it became inevitable to send envoys to Europe. Under these circumstances, curiosity became a state policy and Mehmet Çelebi was nominated the first Ottoman ambassador to Paris.[15]

Mehmet Çelebi was sent to France in order to "visit the fortresses, factories, and the works of French civilisation" and to report on anything that "might be applicable (in Turkey)".[16] In addition to these diplomatic duties, Mehmet Çelebi was implicitly expected to maintain a high level of curiosity to thoroughly investigate and present his findings to the sultan upon his return.[17] During his stay in France, he made observations on art, medicine, botany, zoology, medicine, astronomy, the system of waterways, the position of women in society, the administrative system, architecture and the daily routines of the people. His observations led to the establishment of the printing press, to the construction of palaces and mansions with Western garden designs, and to the emulation of the Western concept of entertainment in the Ottoman Empire.[18]

On the morning of October 7 in 1720, Mehmet Çelebi departed from Istanbul as Ottoman envoy with more than eighty people in his entourage and arrived in Paris five months later (16 March 1721).[19] Being a poet as well as a diplomat, Çelebi occasionally expresses himself in Persian verse in his *Sefaretname*. This was the case, for example, when his ship was stranded by a storm at the beginning of the journey:

> I cried out countless times, no one replied nearby,
> As if no one else exists, under the turquoise dome of the sky.[20]

As they waited on the stranded ship, it began to lurch and sway, its bow and stern repeatedly hitting the ground. The envoy and his retinue were seized with fear and panic, believing that the ship would be torn apart. However, they were saved "*with the help of angels*", miraculously.[21] In line with this description of the storm, Çelebi's *Sefaretname* brims with references to uncertainty, to surprises and to the unfamiliar and the unexpected.

Weeks after their departure, they arrived in Toulon on November 22, "by the grace and mercy of God".[22] News about the ambassador's impending arrival in Toulon had already been dispatched from Paris, and the town's dignitaries eagerly awaited the arrival of the Ottoman delegation. Due to an outbreak of plague, Çelebi's entourage had to spend three weeks in quarantine in Toulon and again for forty days in an abandoned church on the island of Maguelone near Montpellier, before continuing to Paris. In such circumstances, Çelebi "could find no better solution than patience".[23] When they arrived in Sète, a French official welcomed the Ottomans by declaring that "it is certain that this embassy will strengthen the ties of friendship and intimacy that have long existed between the two states".[24]

According to the Cambridge Dictionary, wonder means "a feeling of great surprise and admiration caused by seeing or experiencing something that is strange and new".[25] Without a doubt, the encounter between the Ottomans and the French sparked a sense of wonder on both sides. This, in turn, fed curiosity and the desire to know. Curiosity opens the door to reactive modes such as wonder, comparison, approval or condemnation, and, in the context of travel, the object of curiosity becomes alienated from or compared with the culture of the observer. In this sense, wonder stems from the tension that is built around the duality between the observer and the Other. As soon as Çelebi's delegation arrived in France, the French were affected by a collective emotional response to the strangeness of the Ottomans and, driven by curiosity, gathered on the streets in the hope of catching a glimpse of them. Describing the delegation's arrival in Sète, Çelebi writes that

> The soldiers and captains with their weapons were lined up in two rows until the palace we were going to. It is impossible to describe the crowd here, especially the women... soon the women began to arrive in groups of ten or fifteen. All the rich and esteemed people from the neighbourhood, especially from Montpellier, came with their wives to see us.[26]

As posited by George Simmel, the act of observing the Other with the eye plays a pivotal role in the formation of human social relations.[27] Two groups that do not share certain social and cultural characteristics become objects of tension and strangeness to each other while the strangeness takes a sociological experiential form through visual observation.[28] The capacity of the eye to allow such sociological practice is fundamental to the theme of travel and wonder. Writing about his journey to the Ottoman lands, Henry Blount states that eyes have a more powerful effect than the other senses because they make contact faster and instantaneously: "for above all other senses, the eye having the most immediate, and quicke commerce with the soule, gives it a more smart touch then the rest".[29] The traveller observes events with his eyes and can acquire through a primary experience that is more natural to him. Visual observation—seeing with one's own eyes—provides knowledge that frees the traveller from old interpretations and habits and enables him/her to evaluate what is being seen with a new and sincere feeling.[30] The French wanted to reinterpret their old, imaginative representations of the Turks by seeing Mehmet Çelebi and his retinue with their own eyes. Because, as pointed out by Hughes-Warrington, "we should trust only knowledge seen with our own eyes".[31] The French acknowledged the envoy and his retinue as representatives of Ottoman society, as representatives of different cultural, moral, social and religious norms and made extraordinary efforts to see them. Some Frenchwomen even tried to blend into the Ottoman delegation in disguise.[32] In certain locations, people would undertake journeys of four or five hours merely to observe the Ottomans. People lined up along canals to see the Other with their own eyes. In the jostle and scuttle, some individuals fell in the water, and Çelebi describes how their arrival in Paris caused a riot, which could only be appeased by armed soldiers.[33]

Emphasising the sociological importance of the eye, George Simmel points out that the face carries everything that an individual brings with him about his own life. The clothes, appearance and face of the Ottoman envoy made him a representative of the culture from which he came. Eager to familiarise themselves with the Ottoman Empire manifested in Mehmed Çelebi's face and clothes, the French sought to satisfy their curiosity by means of visual observation.

In *Sefaretname* we discern the utility of wonder as a significant instrument among individuals. Being a historical expedition report that sheds light on the technical and intellectual world of 1720 France and the Ottoman Empire, Mehmet Çelebi's *Sefaretname* offers invaluable insights into the era in which it was created. However, the document is also a literary text that informs readers about early eighteenth-century emotions and culture with references to the interaction between curiosity, travel and wonder. Çelebi's travelogue describes in detail numerous palaces, gardens, exhibition venues and works of art. In so doing, his writing reveals curiosity, which matches the one felt by the French towards the Ottomans.

In a cognitive passion such as curiosity, the notions of strangeness and otherness serve as auxiliary concepts that trigger the desire to know. While "wonder can be triggered by the familiar" as well as by "the strange", the Other is usually positioned in the role of an object about which little or nothing is known.[34] The social relationship between the host and the Other determines what is coded as strange, marginal and exotic. In the era of emerging nation-states of the West, Çelebi represented a multicultural, Oriental Empire. His presence in Paris triggered curiosity and a desire among the French to explore and verify their stereotypical notions and representations of the Ottoman, while also being eager to revise at least some of these stereotypes. As Edward Said argues in his *Orientalism*, it was through stereotypes that the West gradually came to perceive the East as the Other. According to him, the Orient "is a system of representations framed by a whole set of forces that brought the Orient into Western learning, Western consciousness, and later, Western Empire".[35] Although Mehmet Çelebi does not explicitly elaborate on the distinction between the East and the West in the *Sefaretname*, his account nevertheless makes it evident that the French were interested in the Ottoman delegation because of their exotic, Oriental character:

> The people, both men, and women, formed such large crowds on the roads of the cities (we visited) that I often thought that the whole population of our destination had gathered there, leaving no one behind the city. When we entered the house where we were to stay, the onslaught of people who wanted to see us was so great that even the soldiers could not prevent them from rushing in. From time to time, there were even those who got stuck in the chaos and wailed. Some women would come to us unconscious. After leaving us, despite all the trouble they had gone to, they did not go straight home, but waited in the garden in the hope that they might have the

> opportunity to see us again. I personally paid attention to some of them and saw them enter our presence three or four times with great difficulty. In rainy and cold weather, they would wait outside in the courtyard until three or four o'clock at night. We were amazed at their ambition.[36]

For centuries, the French had maintained commercial relations with the Ottomans and therefore had some knowledge about their culture, but for the wider French public a close encounter with the Ottoman delegation was a novelty.[37] As already mentioned, from the Middle Ages onwards, travellers, bureaucrats, merchants, captives and pilgrims travelling from France to the Ottoman Empire or to the East often recorded their observations of cultural differences in order to secure information in writing, and to disseminate their knowledge to Europe.[38] Although travel accounts written by Europeans are occasionally biased and prejudiced, they nevertheless contain observations and knowledge arising from their curiosity about the Orient and Islam. Works such as *The Refutation of the Koran* by Riccoldo da Monte di Croce, a Dominican friar who penned his knowledge of Eastern countries following his travels to the East in the thirteenth century, were reprinted in Paris in 1509;[39] Georgius de Hungary's account of the Turks, initially published in Rome in 1480 and subsequently reprinted in Paris, among others, indicates a strong interest in the *East* in Paris. The works reprinted in Paris and translated into French, as well as the works of those who travelled from France to the East, enriched travel literature. Jean Chardin, who was born in Paris in 1643 and travelled to the East, including the Ottoman lands, in his thirties, describes the differences between the Islamic East and Christian Europe in his *Journal du Voyage du Chevalier Chardin en Perse et aux Indes Orientales*, published in 1686. Fifteenth-century French traveller Gilles de Bouvier et Berry, who took an important role in the King's entourage in 1402 when he was sixteen years old, mentions the way Muslims lived in Ottoman territories.[40] A French soldier, Jean II le Meingre (also known as Boucicaut), captured by the Ottoman Sultan in 1396,[41] discusses the historical context of Ottomans in his account "*Le Livre des faits du bon messire Jean le Maingre, dit Boucicaut*".[42] The French priest and explorer André Thevet (1516–1590), who travelled to the East, recounts his observations in his *Cosmographie Universelle*, which were fuelled by his curiosity about Ottoman society and Muslims. Fresne Canaye explored Istanbul for leisure travel in 1573 and reported his detailed observations in his travelogue. The French polyglot and orientalist Guillaume Postel, who translated the Koran into Latin, wrote several works on Islamic and oriental literature, fuelled by his curiosity. In 1535, he travelled to the Ottoman Empire with a French embassy delegation and documented the information he acquired during his time in the Ottoman Empire in his work *De la République des Turcs*. The French traveller Jean de Thévenot, who followed his curiosity and made travelling a profession, recorded the information he collected about the Ottoman Empire in his

Voyages de M. De Thévenot en Europe, Asie et Afrique. Thévenot, who stayed in Istanbul for nine months from December 1655 onwards, describes the customs and traditions of the Turks, the structure of the cities, and historical and administrative information in his book.

Travel documents from this period relate not only to diplomatic and leisure travel, but also include scholarly studies which combine research and travelling.[43] Antoine Galland's *Journal d'Antoine Galland pendant son séjour à Constantinople: 1672–1673*, consisting of his travel notes published after his travels to the East, contains important information about the Turks. However, it was with the publication of *Les Mille et Une Nuits, Contes Arabes, Traduits en Français* (1704) that Muslims became better known in Europe. There was no similar curiosity for discovery among the Ottomans, who seldom travelled to Europe. In line with this characteristic lack of curiosity, a sixteenth-century Grand Vezir was "quite amazed by the curiosity of the French, who came from so far away for their pleasure without any special or urgent reason".[44]

Çelebi's embassy mission was an important event as it provided the French people with an opportunity to confirm or challenge the centuries-old French imagination regarding the Muslim, the Ottoman and the Turk. As the envoy and his delegation paraded their presence, representing the Ottoman Empire and the image of the East in the minds of Westerners, through various cities of France, they were exoticised by the locals who wanted to detect cultural and visual differences. The ambassador said that it was difficult to describe in words the crowds that came to visit them day and night in the house where they were staying near Paris. Similarly, when they arrived at the King's audience in Paris, people lined up in numerous rows to watch them. Çelebi was aware of the fact that he represented the East and the Ottoman Empire, and that such a position came with certain responsibilities. "Since the people had not actually seen the Ottomans", Çelebi writes, "they had gathered such a crowd just out of curiosity to see what kind of men they were".[45] Later in his travelogue he writes that

> Because the people of Paris had never seen a Muslim in their lives, they did not know the Ottoman attire, hence they were admiring us. The King, they say, remained delighted when they told him about our attire and behaviour.[46]

Çelebi suggests that as soon as the "Eastern other" represented itself to the French, it was viewed with admiration. In these encounters, however, the hosts looked at (and admired) themselves through the figures of the exotic Ottomans who differed from the norm. The expression "they admired us" in Çelebi's statement reflects a mood that resulted from Europeans' past encounters with the East and from their characteristic reactions to otherness. While the French were influenced by the interaction with Çelebi's embassy, the effects were short-lived.

Religion was an important means to perceive and establish otherness. The governor of the Languedoc province (who had probably never witnessed the daily life of an Ottoman) writes that Çelebi continued to fulfil his religious duties during his quarantine, praying five times a day, but drinking wine in a way that one would not have expected from a Turk faithful to the Koran.[47] Besides religion, clothing was a key element in the process of constructing the exoticised Other. National attire of foreigners aroused wonder and facilitated the identification of cultural differences. Parisians travelled long distances to see the costumes of the Ottomans and to witness the cultural representation of the imperial identity, which was embodied in the attire of the Ottoman envoy. Clothing is often taken for granted, but "one person's ordinary can be another person's extraordinary".[48]

After the conquest of Constantinople and the ascension of Suleiman the Magnificent to the throne in 1520, the Ottoman Empire became a power to be reckoned with. It has been argued that "throughout the sixteenth and seventeenth century, the Turk was the object of persistent curiosity on the part of the French public".[49] The French reading public reinforced their notion of the Ottoman through sources such as Pierre Belon's "*Observations de plusieurs singularitez et choses memorables trouvé en Grece, Asie, Judée, Egypte, Arabie* (1553) and Guillaume Postel's '*De la Republique des Turcs*'" (1560). While the Ottoman Empire was a multi-ethnic state, the early modern Western stereotype of the Turk was a crude generalisation, which lumped all ethnic groups together. Çelebi and his retinue were instrumental in validating the representation in order to determine to what extent the French stereotypes of the Ottomans matched the reality.[50] In this process of (re)constructing the Turk, even how an Ottoman ate became an object of curiosity and wonder, as we can judge from Çelebi's account:

> Again, men and women came in large groups, some to visit us and some to watch us; they were especially curious and wanted to see how we were eating. Various questions were coming in like "They are the daughter of so-and-so or the wife of so-and-so, and they are asking your permission to watch you eat". We couldn't refuse many of them, we had to let them come to us. As it coincided with their fast, they did not eat themselves but stood by the table to watch us eat. It was quite challenging for us as we were not accustomed to such a situation; we were only patient to avoid causing offense or upset.[51]

In the scene in which the French wanted to watch the Muslims eat, curiosity and wonder verge on the absurd. The French were equally curious about the religious practices of the Ottomans, observing them perform the tarawih prayer and other religious rituals. In these processes of acquiring knowledge about the world by observing exotic visitors, the members of the host society defined themselves through the Other and, to some extent, moulded their image of the Other to resemble themselves.[52] As D'Aubigny notes, at Çelebi's dinner in Toulouse with the Princess of Auvergne and the president of Riquet, the ladies declared that they saw in Mehmet Çelebi all the refinements of a Frenchman during the meal.[53]

According to Haydn Williams, Çelebi was so courteous and graceful in his demeanour that he changed the centuries-old perception of the Turks as people with less refined manners.[54] This was also implied by Marquis de Bonac, the French ambassador in Istanbul from 1716 to 1724, when he asserted that Çelebi "has more politeness than one should naturally expect from people of this country, and he backs up the politeness of his manners with his speeches".[55] According to the French ambassador, Çelebi was capable of breaking down prejudices against the Easterners. As Williams wryly remarks, there could be no greater compliment than comparing someone's manners to those of a Frenchman.[56] It is evident that Mehmet Çelebi acknowledged his position as the Other and adjusted his behaviour accordingly. Aware of the curiosity directed at him and his retinue, Çelebi repeatedly—albeit reluctantly—allowed the French to watch them eat. The duchess of Orléans Elizabeth Charlotte wrote to the queen of Prussia that Çelebi's politeness made her think that the manners in France were in decline: "judging from the behaviour of the Turkish ambassador here, I am of the opinion that good manners have left the court of France and gone to Turkey".[57]

The letter implies that there was a categorisation of the East and the West, which was based on conceptions of etiquette and manners. These conceptions were then used in the process of recognising and establishing the Other. In his *Passions of the Soul*, Descartes had placed wonder in the first place in his identification of the six primitive passions. When wonder transcends the individual and becomes embodied in a nation, it becomes a subject of historiography, sociology, cultural theory and crowd psychology. Furthermore, when wonder is studied in relation to collective behaviour, it can be seen as an indicator which demonstrates that people who participate in political, industrial and work life also participate in a culture of emotions. The emotions that develop through collective human sociality require "mutual awareness, mutual intentions, and reciprocity".[58] Since humans are social beings, their emotions too are social.[59] When the French observed Çelebi in order to learn about the Ottoman Empire, they were affected by a shared, collective sense of wonder, a conspicuous, single feeling towards the exotic and the Other. In such collective experiences, it has been suggested, "the crowds become like the individual" with "feelings" which in turn "become a form of social presence rather than self-experience".[60]

While emotions contribute to the establishment of hierarchical structures, the place of the Orient in the old hierarchy between the East and the West began to change due to Mehmet Çelebi's stay in Paris. It has been argued that "his gentle personality, intelligence, harmonious and courteous behaviour played a role in changing the image of Turks in Paris".[61] Çelebi's charm stemmed from his diplomatic skills and polite manners, and from his practice of spoiling French ladies with jam, coffee and sherbet.[62] This was a significant change in comparison with the past perceptions on both sides.

A Journey into Wonder: Mehmet Çelebi's Impressions

Since the Ottoman Empire considered itself superior to the Christian West, it remained for a long time unaware of technical developments of the West. For instance, Georgius de Hungaria, who lived as a slave in Ottoman lands between 1436 and 1458, made the following observation about the *Turks*: "For none of the liberal arts appear to be taught in their schools. They are so foreign from all other sciences, that the names of different disciplines are not found or recognised by them".[63] Mehmet II, the conqueror of Istanbul, kept up with advances in the West, but as the more Eastern lands were conquered and the Ottoman Sultan became caliph, the growing influence of Islam caused the Ottoman Empire to close itself from the West.[64] Montesquieu, who published his epistolary novel "Persian letters" about the East and the West in the same year when Mehmet Çelebi was in Paris (1721), takes his readers to a number of Ottoman cities and, through the voices of his protagonists, demonstrates how far behind the Ottomans were in terms of technology. "While the nations of Europe advance further every day", one of his characters declare, "Ottomans have paid so little attention to technical knowledge and the new inventions of the European nations".[65]

Mehmet Çelebi's journey to Paris was followed by efforts to westernise the Ottoman Empire. These efforts intensified in the nineteenth century, and sending students to Paris became one of the new cultural policies of the state. This led to the spread of Western ideas, literature and political attitudes. Paris played an important role in the modernisation of the Ottoman Empire and, subsequently, the Republic of Turkey.[66] The significance of Paris was emphasised by the scholar and future rector of Istanbul University Hoca Tahsin Efendi (who was sent to Paris in 1856 to assist Ottoman students) when he declared that "travel to Paris, if wisdom is your desire".[67]

Being one of the first Ottomans to travel to Paris, Mehmet Çelebi was impressed by the experience. His interest in technical matters contributed towards bridging the gap between the Ottoman Empire and the West. At the same time, the *Sefaretname* pays less attention to the customs, rituals and daily life of the French.[68] In Paris, he observed developments especially in botany, architecture, pharmacy, hospitals, astronomy and art. Rather than an embassy report, *Sefaretname* is a sort of wandering narrative in which Çelebi constantly discovers new things as curiosity drives him to places such as mirror workshops, palatial gardens, a Gobelin tapestry workhouse and an observatory.

Çelebi's travelogue makes comparisons between Christianity and Islam, East and West, the Ottoman and the Other. The importance of *Sefaretname* is acknowledged by Ahmet Hamdi Tanpınar when he writes that "no book can occupy such an important place in our history of Westernisation as this little report". Furthermore, "the idea of a comparison is hidden in almost every line and contains almost the entire programme of subsequent changes".[69] Although Çelebi does not separate East and West as sharply as often happens in today's narratives, his text nevertheless distinguishes between his own perspective and that of the French. However, this distinction does not isolate one from the other, but allows mutual interaction. Following Çelebi's stay in Paris, the Ottoman style in clothing and decor called Turquerie became fashionable in France, while Istanbul welcomed French influences in style and taste.[70]

On his journey to Paris, Mehmet Çelebi was a representative of his own cultural background, although he travelled with a strong intention to leave his prejudices behind. Motivated by curiosity, he was receptive to the wonders he encountered.[71] When he visited the Gobelin workhouse, the workers hung all the carpets they had woven on the walls in anticipation of his arrival. Çelebi could not hide his astonishment at the human faces woven on Gobelin carpets, because, as it has been pointed out, the Ottomans were not familiar with the European traditions in plastic arts and painting.[72] Even the most basic elements of the textile art aroused in him the astonishment of a child:

> When we watched the carpets hanging on the walls, we were speechless with amazement. For example, they have woven flowers, when you look at them, you think they are alive in jars. In their depictions, the eyes, eyelashes, eyebrows, and especially the hair and beard are so vividly embroidered that even Mani and Bizhad would not be able to perform so well on Chinese paper. Some of the figures are shown with smiling faces to convey joy, some with sad faces to convey sorrow, and some with tears to convey fear. So the facial expressions in the pictures stood out at a glance. No matter how much we try to explain, the images themselves are very different from what we describe.[73]

In his definition of wonder, Descartes writes that "when the first encounter with some object surprises us, and we judge it to be new, or very different from what we knew in the past or what we supposed it was going to be, this makes us wonder and be astonished at it".[74] Mehmet Çelebi was astonished by the portrayals on the carpets because Islamic restrictions had prevented the development of this art in the Ottoman Empire. For example, according to what Mehmet Çelebi recorded in his *Sefaretname*, typically, the king would have portraits of himself made, adorned, and given to the envoys. But, "since portraying a face is not permissible in Islam, he gives alternative objects instead".[75] According to the Turkish poet Yahya Kemal Beyatlı, "we cannot see the faces of our ancestors because of the lack of painting".[76] He argues that because in the Ottoman Empire the art of painting was restricted from the beginning, the Ottomans were confined to a limited sphere in terms of understanding life and human beings. Fascinated by the portrayal of human faces on the carpets, Çelebi identified the representation of sorrow in a sad face and joy in a smiling face as a new kind of skill or knowledge, which made a connection between the emotions and drawing. Çelebi believed that it was impossible to adequately convey his astonishment, as he was convinced that words were inadequate to fully capture his profound sense of wonder. In line with his conviction about the limits of language, he occasionally refrained from describing in detail the innovations and objects of wonder that he encountered.

Çelebi was also impressed by opera, another art form that did not exist in his own culture. He writes that "there was an artistic show peculiar to the city of Paris, called opera, they would show strange arts, and there would be a large crowd. We watched it".[77] He describes the opera in detail, providing information on dance, costume, light and decor besides explaining the interiors of the venue, the organisation, the work that was being staged and the appearance of the audience. Attentive to every detail, he even describes the lighting of the candles in the hall. He was surprised at the sudden changes in scenes and decorations during the performance. For instance, he describes a scene in which a palace is suddenly replaced by lemon and orange trees, allowing a princess to enter a garden. Then, when the princess starts to pray, the garden is quickly replaced by a large church. All this leaves him so astonished that, again, he considers it difficult to describe: "in short, they showed so many things that surprised people that it is impossible for me to describe them all here".[78]

Wonder is a state of mind that indicates we have reached the limits of our comprehension.[79] It seems that this was what Mehmet Çelebi experienced at the opera: when describing his experience, he uses the word *hayret* (derived from the Arabic root "*hyr*"), which means wonder. His astonishment and positive appraisal of opera as an art form is further manifested in the adjectives he uses to describe the experience, according to him, which conveyed stories and emotions through the creation of a complex and harmonious composition. He states that "the opera was ostentatiously built and the banisters, columns, walls and ceilings were decorated with gold" and that, in opera, the objective "is to convey stories through live performance. Each story is published as a book, thirty books alltogether. Each story has its own title, and in every gathering, they present a story as if it is newly unfolding". In opera, "they showed such wonderful things that they were beyond words" and, Çelebi adds, "unusual things were witnessed". For example, he writes, "the area designated for dancing was filled with fairy-faced beauties, and an incredibly bright sun appeared as if it had just risen. The sun, wide as a large tray, was crafted with such skill in gold that the light from the candles behind it shone like real sunlight".[80] Afterwards, Çelebi introduced opera to the Ottoman Empire where, as a part of the project of adopting Western cultural practices, it became a prominent art form and remained such also in the Republic of Turkey.

The fact that women and men sat next to each other at the opera was a novelty for Çelebi, who came from a culture in which women did not participate in social life. The women who passed "in front of the candle flames to take their seats" while wearing "gold and silver-embroidered fabrics and jewellery" was, according to Çelebi, "such an astonishing sight" that it was impossible to describe.[81] He marvelled at the "dancers in special costumes" and the "twenty fairy-like figures", who appeared "in jewelled costumes to add glamour to the gathering".[82] In the Ottoman Empire women were not allowed to walk freely in the streets, let alone dance or sing in public. These restrictions resulted in the so-called *zenne-köçek* culture in which men dressed as women on stage and performed dance and theatre. In this light, it is hardly a surprise that throughout the *Sefaretname* Çelebi marvels at the position of women in the French society. Similarly, Ahmet Resmî Efendi was surprised by the freedom of women during his embassy mission to Vienna (1757–1758), where he marvelled at the unrestricted presence of women at events such as the opera and ball.[83] Discussing the position of women in Paris, Çelebi writes that

> in France, women are held in much higher esteem than men. The women can do whatever they want and go whereever they desire, and no one can object. To the lowest of them, the finest gentlemen would demonstrate an excess of regard and respect. In these lands, the women's word is accorded greater respect and influence. So much so that they called France a women's paradise. Indeed, in these lands, women are not subjected to almost any hardship or trouble. All their wishes and desires were fulfilled immediately without any delay.[84]

Given the freedom of women in the French society, women in Paris displayed a great interest in Mehmet Çelebi and his retinue. Çelebi was struck not only by the sociability of women but also by the keen interest they showed in a man. French women found the Ottomans exotic and, every night until dawn during Ramadan, admiringly observed them perform the tarawih prayer and dine. Çelebi was equally curious about French women and once politely requested some of them to remove their gloves so that he could admire their hands.[85] It is unlikely that he would have made such a request to an Ottoman woman in his own culture.

During his journey, Çelebi's sense of wonder and astonishment intensified to the point at which he started to believe that those who had not experienced the same would not be able to comprehend his written account. When he could not find adequate words to describe a wonder in front of him, he resorted to terms such as "uncountable", "inexpressible" and "indescribable", surrendering to the limited expressive power of language, which "is not everything in emotion".[86] This suggests that he did not wish to impose limitations on the emotions he had felt—wonder, astonishment, awe—by trying to define them with words. As Martha Nussbaum has pointed out, "in the process of labeling, we are also frequently organizing, bounding something off from others, sharpening distinctions that may have been experienced in an inchoate way. From then on, we experience our emotions in ways guided by these descriptions".[87]

Conclusion

To conclude, it is evident that travelling exposes one to the unknown, to cultural identities and moral values which differ from the familiar and therefore have the power to bewilder and astonish. Ahmet Ağaoğlu (1869–1939), who studied at the Sorbonne and later became one of the intellectuals of Ottoman Istanbul, elaborates on the stark differences between European and Islamic civilisations in his work *Üç Medeniyet* (Three Civilisations). He asserts that if one were to leave Istanbul and travel to Marakesh, Al-Jazira and Cairo, one would not experience a sense of foreignness because one would encounter the same turbans and similar status for women, a familiarity which would eliminate any sense of wonder. Similarly, he posits that an individual from Rome would feel at home in Paris, Vienna and Berlin, because the mood, way of thinking, way of life and religion are all similar. Every civilisation has a common way of thinking and feeling, and therefore a shared sense of morality and taste.[88] In Paris, Çelebi lacked familiar surroundings, mindset and habits. The profound cultural differences that he perceived during his journey filled him with wonder and the feeling of being immersed in a different world. His sojourn in France made him conclude that "so many unfamiliar *(tuhaf)* things have occurred that is imposible to enumerate them all".[89] What is more, he recalled the *hadith* of the Prophet Muhammad: "this world is the dungeon of the Muslim and the paradise of the infidel".[90] Çelebi's *Sefaretname* recounts an adventure of gaining new insights and, in so doing, demonstrates the transformative quality of travel and wonder.

Notes

1 Yirmisekiz Mehmet Çelebi, *Sefaretname-i Fransa*, 28.
2 See, for example, D'Aubigny 1889, 78–91, 200–235; Fatma Göçek Müge 1987; Lunde 1993, 30–39. Kılıç, 2003, has also analysed the *Sefaretname* regarding Paris as a scientific centre in the context of the travel of science, examining the dissemination of knowledge.
3 Marr 2016, xv.
4 Blount, *A Voyage into the Levant*, 1.
5 Blount, *A Voyage into the Levant*, 2.
6 In their travelogues, Western writers often refer to the Ottoman Empire as the Turkish Empire and to the Ottoman inhabitants as "Turks" regardless of their ethnicity.
7 See, Fügedi, Erik (ed), *Kimondhatatlan nyomorúság*.

8 Schiltberger, *Hans Schiltbergers Reisebuch*; Akpınar 1985, 353.
9 Said 1979.
10 Çelebi, *Seyahatnamesi*, 12.
11 Berkes 1998, 30.
12 Tanpınar 1988, 43.
13 Western states established permanent embassies in the Ottoman Empire from the fifteenth century onwards, for example, Venice 1454, Poland 1475, Russia 1495, France 1525, England 1583 (for more on this, see Unat 1987.
14 Girgin 1994, 40; Berkes 1998, 29.
15 D'Aubigny 1889, 83.
16 Berkes 1998, p. 33.
17 For more on this, see, D'Aubigny 1889, 78–91.
18 Lewis 2002, 45–50.
19 Akyavaş 1993, viii.
20 Çelebi, *Sefaretname*, 3:*"Feryâd besî kerdem ve feryâd resî nist.*
Güya ki der in künbed-i firuze kesî nist".
21 Çelebi, *Sefaretname*, 3.
22 Çelebi, *Sefaretname*, 5.
23 Çelebi, *Sefaretname*, 9.
24 Çelebi, *Sefaretname*, 11.
25 *Cambridge Dictionary*, *s.v.* "wonder".
26 Çelebi, *Sefaretname*, 12.
27 Simmel 1969, 358.
28 Simmel 1969, 358.
29 Blount, *A Voyage into the Levant*, 3.
30 Blount, *A Voyage into the Levant*, 3.
31 Hughes-Warrington 2019, 47.
32 Veinstein 1981, 47.
33 Çelebi, *Sefaretname*, 15.
34 Hughes-Warrington 2019, 8.
35 Said 1979, 202–203.
36 Çelebi, *Sefaretname*, 22.
37 D'Aubigny states that the people from the East had a bad reputation at Versailles. Suleyman Aga, previously sent to Paris in 1669, caused absurd scenes, and an unsuccessful embassy experience. (D'Aubigny 1889, 80–83.).
38 For more on this, see Örik 1929.
39 Fügedi, Erik, (ed.), *Kimondhatatlan nyomorúság*, 210.
40 Örik 1929, 64.
41 Pears. 1923, 676.
42 Örik 1929, 64.
43 Çiğdemoğlu 1998, 109.
44 When Pierre Lescalopier, a lawyer in Paris, came to Istanbul and had an audience with Sokollu Mehmed Paşa, the latter said in amazement: "For a Frenchman to come from so far away for no reason in particular is a sign of the French taking delight in such things" (Örik 1929, 71).
45 Çelebi, *Sefaretname*, 28.
46 Çelebi, *Sefaretname*, 37.
47 D'Aubigny 1889, 86.
48 Hughes-Warrington 2019, 12.
49 Keller 2013, 2.
50 Lippman 1922.
51 Çelebi, *Sefaretname*, 29.
52 Said 1979.

53 D'Aubigny 1889, 89. On the idea that a host society defines itself through the Other, see Said 1979.
54 Williams 2015, 42.
55 D'Aubigny 1889, 81.
56 Williams 2015, 42.
57 Williams 2015, 42.
58 Nussbaum 2003, 148.
59 Nussbaum 2003, 14.
60 Ahmed 2014, 10.
61 Arıkan 2013, 551.
62 Williams 2015, 42.
63 Stevenson 2016, 85.
64 Veinstein 1981, 10–11.
65 Montesquieu, *Ouvres Complètes, Vol. I, Lettres Persanes*, 98.
66 A century later, a comic Western alafranga character emerged in Turkish literature, based on the stories of those who had travelled to Paris.
67 "*Parise git ey efendi akl u fikrin var ise Âleme gelmiş sayılmaz gitmeyenler Paris'e*".
68 Veinstein 1981, 42.
69 Tanpınar 1988, 44.
70 Sözen 2009, 17–32.
71 Stagl 2004, 4, 11, discusses "curiosity as the first motivation" in the context of travel.
72 Tanpınar 1988, 33.
73 Çelebi, *Sefaretname*, 59–60.
74 Descartes, *The Passions of the Soul*, 52.
75 Çelebi, *Sefaretname*, 77.
76 Beyatlı 1970, 69.
77 Çelebi, *Sefaretname*, 45.
78 Çelebi, *Sefaretname*, 46: "*Netice-i kelâm ol kadar hayret-fezâ şeyler gösterdiler ki*".
79 Opdal 2001, 332.
80 Çelebi, *Sefaretname*, 45–47.
81 Çelebi, *Sefaretname*, 46.
82 Çelebi, *Sefaretname*, 46.
83 Dadan 2006, 276.
84 Çelebi, *Sefaretname*, 12.
85 D'Aubigny 1889, 88. (as cited in a letter from the Duke of Roquelaure on February 28, 1721).
86 Nussbaum 2003, 149.
87 Nussbaum 2003, 149.
88 Ağaoğlu 1927, 4–5.
89 Çelebi, *Sefaretname*, 59.
90 Çelebi, *Sefaretname*, 54: "*Eddünya sicn'ül-mümin ve cennet'ül kâfir*".

Bibliography

Primary sources

Akyavaş, Beynun. *Yirmisekiz Çelebi Mehmed Efendi'nin Fransa Seyahatnamesi*. Ankara: Türk Kültürünü Araştırma Enstitüsü, 1993.

Blount, Sir Henry, *A Voyage into the Levant*. London: I.L. for Andrew Crooke, 1637.

Çelebi, Evliyâ. *Evliyâ Çelebi Seyahatnamesi*. Ed. by R. Dankoff, S.S. Kahraman, Y. Dağlı. Vol I. Istanbul: Yapı Kredi Yayınları, 2006.

Descartes, René. *The Passions of the Soul*. Trans. by. S. H. Voss. Indianapolis/Cambridge: Hackett Publishing Company, 1989.

Fügedi, Erik. Kimondhatatlan nyomorúság. *Két emlékirat a 15-16. Századi oszmán fogsáról*. [Unspeakable Misery. Two memoirs about the Ottoman captivity in the 15-16th centuries] Trans. and ed. by Erik Fügedi. Budapest: Európa, 1976.

Montesquieu. *Ouvres Complètes de Montesquieu, Volume I. Lettres Persanes*. Ed. by Edouard Laboulaye. Paris: Garnier Frères, 1875.

Schiltberger, Johannes. *Hans Schiltbergers Reisebuch*. Tübingen: H. Laupp, 1885.

Yirmisekiz, Mehmet Çelebi. *Sefaretname-i Fransa*. Istanbul: Matbaa-i İlmiyye-i Osmaniyye, 1283 (1866).

Literature

Ağaoğlu, Ahmet. *Üç Medeniyyet*. Istabul: Türk Ocakları Merkez Heyeti Matbaası, 1927.

Ahmed, Sara. *The Cultural Politics of Emotion*. Edinburgh: Edinburgh University Press, 2014.

Akpınar, Turgut. Almanların Marco Polo'su Schiltberger (1380-1440) ve Eseri. *Tarih ve Toplum*, c.3 (1985): 353–355.

Arıkan, Zeki. "Yirmisekiz Çelebi Mehmed Efendi" *TDV Islam Ansiklopedisi*, C 43, Ankara: TDV (2013): 551–552.

Berkes, Niyazi. *The Development of Secularism in Turkey*. New York: Routledge, 1998.

Beyatlı, Yahya Kemal. *Edebiyata Dair*. Istanbul: Istanbul Fetih Cemiyeti Press, 1970.

Çiğdemoğlu, Sema. "Orta Çağ, XVI., XVII., XVIII. Yüzyıllarda Fransız Doğu Seyyahlarına Kısa Bir Bakış". *Ankara Üniversitesi Dil ve Tarif-Coğrafya Fakültesi Dergisi*. Volume 38, Issue 1-2, (1998): 103–120.

Dadan, Ali. "XVIII. Yüzyılda Osmanlı Sefirlerinin Avrupa Algısı (Kâfirlerin Cenneti)". *Marife*. 3 (2006): 269–281.

D'Aubigny, E., "Un Ambassadeur Turc à Paris sous la Régence". *Revue d'Histoire Diplomatique*, III (1889): 78–91, 200–235.

Göçek, Fatma Müge. *East Encounters West: France and Ottoman Empire in the Eighteenth Century*. New York: Oxford University Press, 1987.

Girgin, Kemal. *Osmanlı ve Cumhuriyet Dönemlerinde Hariciye Tarihimiz (Protokol ve Teşkilat)*. Ankara: Türk Tarih Kurumu, 1994.

Hughes-Warrington, M. *History as a Wonder: Beginning with Historiography*. London: Routledge, 2019.

Keller, Marcus. "The Turk of Early Modern France", *L'Esprit Créatur*, Vol. 53, No. 4 (2013): 1–8.

Kılıç, B. "Yirmisekiz Mehmed Çelebi's Travelouge and the Wonders that Make a Scientific Centre. In *Travel of Learning*, Boston Studies in the Philosophy of Science, Vol 233, edited by A. Simões, A. Carnerio, M.P. Diogo, 77–100. Dordrecht: Springer, 2003.

Lewis, Bernard. *The Emergence of Modern Turkey*. New York & Oxford: Oxford University Press, 2002.

Lippman, Walter. *Public Opinion*. New York: MacMillan Co., 1922.

Lunde, Paul. "A Turk at Versaille", *Aramco World* Vol. 44, No 6 (1993): 30–39.

Marr, Alexander. "Introduction". In *Curiosity and Wonder from the Renaissance to the Enlightenment*, edited by R.J.W. Evans and A. Marr, 1–20. Oxon & New York: Routledge, 2016.

Nussbaum, Martha. *Upheavals of Thought: The Intelligence of Emotions*. Cambridge: Cambridge University Press, 2003.

Opdal, P.M. "Curiosity, Wonder, and Education Seen as Perspective Development". *Studies in Philosoph and Education* 20 (2001): 331–344.

Örik, Nahid Sırrı. "Şarkı Karipteki Fransız Seyyahları Hakkında Bir Eser". *Türk Tarih Encümeni Mecmuası*, Vol. 1 (1929): 63–75.

Pears, Edwin. "The Ottoman Turks to the Fall of Constantinople". In *The Cambridge Medieval History, Vol IV, The Eastern Roman Empire (717-1543)*, edited by J.R. Tanner, W. Previte-Orton, Z.N. Brooke, 653–705. Cambridge: Cambridge University Press, 1923.

Said, Edward. *Orientalism*. New York: Vintage Books, 1979.

Simmel, G. "Sociology of the Senses: Visual Interaction". In *Introduction to the Science of Sociology*, edited by Robert E. Park and Ernest W. Burgess, 356–361. Chicago: Chicago University Press, 1969.

Sözen, Kemal. XVIII. Yüzyıl Osmanlı-Batı İlişkisinin Yazılı Vesikası: Paris Sefaretnamesi ve Batılılaşma Eğilimi Açısından Değerlendirilmesi, SDÜ İFD, 2009: 17–32.

Stagl, Justin. *A History of Curiosity: The Theory of Travel 1550-1800*. London and New York: Routledge, 2004.

Stevenson, David Ryan. *An English Translation of Georgius de Hungaria's Treatise on the Customs, Living Conditions, and Wickedness of the Turks (1481)*. Diss., Emory University, 2016.

Tanpınar, A.H. *XIX. Asır Türk Edebiyatı Tarihi*. Istanbul: Çağlayan Press, 1988.

Unat, Faik Reşit. *Osmanlı Sefirleri ve Sefaretnameleri*. Ankara: Türk Tarih Kurumu Yayınları, 1987.

Veinstein, G. "Introduction" in *Le Paradis des Infidèles: Relation de Yirmisekiz Mehmed Efendi, Ambassadeur Ottoman en France sous la Régence*, 7–51. Paris: François Maspero, La Découverte, 1981.

Williams, Haydn. *18. Yüzyılda Avrupa'da Türk Modası: Turquerie*. Trans. by. N. Elhüseyni. Istanbul: Yapı Kredi Yayınları, 2015.

8

WONDERS AND CURIOSITIES

Early modern British impressions of Bohemia

Hana Ferencová

Introduction

Drinking cups in the shape of dragons, an ostrich made of its eggs, Japanese frier carved in box, a dart of a unicorn horn—these are just some of the remarkable items contained in the cabinet of curiosities in Dresden visited by the English traveller Edward Browne on his journey from Vienna to Hamburg in the cold autumn days of the year 1669. When describing the elector's palace in the city, it was the "Kunstkammer, Art-Chamber, or Collection of Rarities, both of Art and Nature", which deserved the greatest attention.[1] In his guide book *The Grand Tour*, published eighty years later, Thomas Nugent extols the collection and declares that the cabinet of curiosities is "the most remarkable thing, and most agreeable to travellers". He then mentions several extraordinary examples of objects contained in the collection.[2]

The sixteenth and seventeenth centuries witnessed a surge of interest in a vast array of artificial and natural objects originating from different parts of the world, amassed and categorised in cabinets of curiosities by European collectors.[3] These exhibits reflected the collectors' perceptions and knowledge of the world beyond the boundaries of the normative, encompassing rare, curious and wondrous items. The tendency to conceptualise nature in terms of the marvellous was a particularly pervasive phenomenon in early modern Europe.[4] As David Jones has pointed out, these collections facilitated global understanding and exploration, enabling readers to gain insight into unfamiliar cultures and experiences and transforming them into known ones. How people interacted with the extraordinary and wonderful and how they presented this experience in travel writing had important consequences. While travel enabled

DOI: 10.4324/9781003506928-8

the individual to gain a sense of the vastness of the world, the act of collecting enabled them to bring that experience and knowledge back home. Collections of rare and exotic objects turned the marvellous and the wondrous—which had thus far remained beyond the scope of human comprehension—into something real and tangible. This in turn had the effect of reducing the sense of wonder associated with such objects, slowly transforming wonder into curiosity in the course of the early modern era.[5]

However, the connection between what travelogues continued to portray as wondrous and what was considered curious is inherently tenuous and not easily definable, particularly in the eighteenth century, when there was a significant shift in the tone of travelogues towards a more scientific recording of the observed countries. This shift correlated with the emergence and popularity of the methodologically oriented apodemic literature.[6] In general, when expressing amazement and astonishment, travel writers used a variety of adjectives ranging from wondrous and marvellous to curious, rare, remarkable, extraordinary, *etc*. This chapter focuses on curiosities and wonders recorded in such English travelogues which were published in the period between the late seventeenth and mid-eighteenth century, that is, in a period during which the genre of travel writing went through significant changes and developments.

Geographically, the chapter explores Central Europe and in particular the Bohemian Lands, which were located in the heart of the European continent, a considerable distance away from the usual Grand Tour routes further southwest. Although the area attracted a number of early modern English travellers, it should be stressed that they considered Bohemian Lands a transit region rather than a final destination and, as a consequence, rarely spent long periods there. Usually, English travellers visited the region on their way from Dresden to imperial cities such as Vienna. Prague, the Bohemian capital, attracted more attention than any other city because it was located on the main route through the region.[7]

In what follows, I focus on the travels of three seventeenth- and eighteenth century Englishmen: Edward Browne, Richard Pococke and John Swinton. All three travellers were educated at Oxford or Cambridge and later set out on a several-years-long Grand Tour to the continent in order to complete their education. In contrast to many of their compatriots, Browne, Pococke and Swinton travelled through Bohemia. The first of these journeys was made by Edward Browne (1644–1708), a prominent physician and author who was also a member of the Royal Society. In the autumn of 1669, he undertook a journey to Bohemia, which he later chronicled in his publication *An Account of Several Travels through a Great Part of Germany* (1677).[8] His travelogue was translated into French and Dutch, and it was published in multiple editions during the eighteenth century.[9] Well received by European readers, it also served as a source of inspiration for Swinton's travel journal.[10]

The second traveller discussed here was John Swinton (1703–1777), a clergyman of the Church of England, writer, orientalist and historian. He passed through Bohemia in 1734, during his journey from Venice to Hamburg. In 1745 and 1746, Swinton's travelogue *The travels of three English gentlemen* was published in three parts in *The Harleian Miscellany* (edited by Samuel Johnson and William Oldys and published by Thomas Osborne).[11] It was reprinted on four further occasions by the beginning of the nineteenth century. For a long time, Swinton's travel journal was regarded as an anonymous text, and the authorship was only established by the historian Trevor Shaw towards the end of the twentieth century.[12]

The third English visitor to Bohemia was Richard Pococke (1704–1765), a bishop, writer and traveller, who is primarily remembered for his narrative about the Middle East.[13] However, he travelled through Bohemia only three years later, in 1737, and recorded this journey in his famous book *A Description of the East and Some Other Countries* (1745). The chapter on Bohemia is relatively brief, comprising no more than four pages.[14] However, the original manuscript, which describes the visit to the Bohemian Lands, is a considerably more substantial document. It comprises 122 folios and is written in the form of a separate travel book. Consequently, it constitutes a valuable source of information that is further enhanced by the inclusion of a number of illustrations that were excluded from the published version.[15]

In harmony with the age of Enlightenment, the selected travelogues all aspired towards an objective and scientific style of writing. However, in accordance with contemporary practice, the writers continued to inform their readers about novelties, curiosities and extraordinary phenomena.[16] For instance, as Kees van Strien remarks, Edward Browne gained international renown as a travel author due in part to his father's influence, who encouraged him to adapt the letters he originally sent home for publication. He advised his son to refrain from excessive focus on scientific details and to provide illustrative accounts of the cultural nuances of each country, suggesting relevant reading on the subject matter. The outcome was a factual and learned mixture of information on what was "Naturally, Artificially, Historically, and Topographically remarkable".[17] The aim of this chapter is to examine the representation of wonders and curiosities in the Bohemian section of Browne's, Pococke's and Swinton's travel journals. In so doing, the chapter explores the tension that arises from the fact that the travelogues were at once adopting a more scientific and objective approach and, like most travel writing, allured by wonders. Furthermore, the chapter considers how these English travellers discussed wonders and supernatural matters in relation to the culturally unfamiliar context of the visited region. Finally, the chapter considers how the authors addressed the problem of credibility and readers' doubts when describing wonders in their travelogues.

Natural curiosities and extraordinary legends

The *Kunstkammer* or the cabinet of curiosities of Prague made accessible artistic and natural examples of rarities that could not be seen in the city at the time of Browne, Swinton and Pococke, because the famous collection of exotic objects of the emperor Rudolf II had been sold after his death in 1612. It was not the case, however, that Bohemian curiosities would have been limited to such collections of rarities. Referring to the spa and mining towns of the region, Richard Pococke declares that "the kingdom of Bohemia abounds in natural curiosities".[18] The eighteenth century witnessed a surge in the popularity of spas in Europe with the consequence that Bohemian spa towns became popular tourist destinations. Of particular importance in this respect was the western and northwestern Bohemian region, where Karlovy Vary (Karslbad) attracted a considerable number of guests from across Europe. One of them was the Russian Tsar Peter I, who stayed in Karlovy Vary in 1711 and again in the following year.

The English travelogues discussed in this chapter provide detailed accounts of the spa treatments and the curative effects of different springs. In so doing, Browne, Swinton and Pococke drew on information derived from a variety of sources, including non-medical books, which can be seen as precursors of later guidebooks.[19] In two cases, I have found almost identical passages describing the health resorts. First, the description of spa treatments in the published version of Richard Pococke's travelogue says: "The mill-spring at some distance from this is of the fame nature, but not so hot, nor so strong of the minerals as the other; it is used for bathing on the spot, and both for bathing and drinking by persons of warm and weak constitutions; the other waters being proper for the cold and robust; these waters in general are good for all obstructions, particularly for the gravel and barrenness".[20] The text is entirely identical to the information contained in Thomas Nugent's guidebook to Germany published four years later so that Pococke's book either inspired him or both authors drew from the same source.[21] Focusing on technical descriptions of spas, springs and available treatments, neither book refers to the miraculous, healing power of such cures. However, the detailed descriptions of the resorts suggest a curiosity typical of a traveller who is exposed to something wondrous.

Like his published travelogue, Pococke's manuscript on Bohemia has a noticeably scientific approach to the origins, effects, quality and colour of mineral water. As far as the natural springs of Karlovy Vary were concerned, he was astonished by the diverse and colourful nature of rocks in the river formed by the springs, which could be seen in this spa town: "What is most extraordinary in the nature of this water is its forming a stone substance when it runs".[22]

Besides the spa towns, Pococke shows interest in the mines of Bohemia, admiring the abundance of precious stones.[23] As already mentioned, in contrast to France and Italy, Bohemia was a transit destination and, as such, most often visited during a more extensive journey from Dresden to Vienna, with a few days in Prague. Although travellers typically spent only a short time in other areas of the country (perhaps with the exception of spa towns in Western Bohemia), they did describe the regions that they visited and occasionally recounted anecdotes that they had heard from the local inhabitants.

One such anecdote relates to the story of Rübezahl, a folkloric mountain spirit of the Giant Mountains (*Krkonoše* in Czech) along the border between the historical lands of Bohemia and Silesia. Rübezahl is the subject of many legends and fairy tales in German, Polish and Czech folklore. The earliest known representation of the mountain spirit can be found on the 1561 map of Silesia by Martin Helwig, where it is depicted as a black-tailed demon. The figure, which is composed of several zoomorphic parts, is situated above the Giant Mountains between the town Kowary and the highest peak Sněžka, with its body leaning on a massive pilgrim's staff.[24] Edward Browne was aware of the legend, as we can judge from a passage in his travelogue. On his journey from Vienna to Prague, Browne met a gentleman who was coming from the Giant Mountains. Curious about the legend, Browne wanted to learn more: "I enquired of him concerning the spirit Ribensal, which is said to infest that Country, but he could say nothing therein of his own knowledge; and though he was confident that there was such a Spirit; yet he confessed that for twelve years it had done no hurt".[25]

Only few years before Browne's journey, the legend of Rübezahl was made known in Europe by Johannes Prætorius' *Dæmonologia Rubinzalii Silesii* (1662). In Czech legends and fairy tales, Rübezahl is known as Krakonoš. Over time, his character evolved from a malevolent demon causing storms and heavy snow to that of a guardian of the impoverished people living in the Giant Mountains. In addition to this, his physical appearance has varied over the centuries, and currently he is depicted as a tall, elderly man with a long beard, a giant capable of crossing his mountains with one step.[26]

In the passage that deals with the legend of Rübezahl, Edward Browne confirms the popularity of the tale: "in Hills, Mountains, and places of Mines such reports are ordinary".[27] He then proceeds to give further examples of spirits that are said to haunt mines across the country. One such haunted place was Kutná Hora (Kuttenberg), a silver mining town close to Prague.[28] In the Middle Ages, Kutná Hora was a prosperous town due to its silver mining activities. However, at the beginning of the early modern period, its importance began to decline. Browne visited Kutná Hora and descended into the oldest mine, which gave the town its name: "it is called the *Cotna*, or *Auf der Gotten*, upon the *Gotten*, or *Goat hill*; and as the Story goeth, a *Monk* walking over this Hill found a kind of a *Silver Tree* sticking to his Garment, which was the occasion that they afterwards digged and built these *Mines*, and the place retains the name of *Gottenberg*".[29] In his description of the place, Browne refers to another legend that was connected to the silver mines of Kutná Hora: "I have read that

Libussa the Princess, and reputed Sorceress of Bohemia, foretold many things concerning these Mines; but certain it is, that for the advantage and profit these bring, the Town hath suffered much in many wars".[30] It is unclear from Browne's description to what extent he believed in the miraculous aspects of the recorded stories. As far as the legend of Princess Libuše was concerned, he sought to verify the legend with references to specific historical facts, describing, for example, how the town was sacked in the early fifteenth century due to its wealth. Browne's treatment of the subject may have been influenced by his learned father, Thomas Browne (1605–82), whose publication *Vulgar Errors* (1646)—balanced between scientific and fictional musings—sought to expose the errors and superstitions of the era.[31]

Similarly to Browne's travelogue, John Swinton's *The travels of three English gentlemen* sought to engage the reader through references to local curiosities, unusual objects and extraordinary experiences such as folk tales of ghosts and spirits. For instance, Swinton refers to a ghost that caused fear amongst those living in the vicinity of the church of St. Peter in Hamburg: "the person, who mentioned this, is a merchant of great substance and reputation. He assured us, that scarce anybody in Hamburg doubted of the reality of it. But, notwithstanding this is so well attested, we leave our readers, as we have done upon a parallel occasion, to believe as much or as little of it as they please".[32]

In the case of Bohemia, there was a story of the so-called White Lady who haunted the chateau in Jindřichův Hradec (Neuhaus), a town located in the southern part of the region. The chateau was constructed by an aged widow who—according to the legend—continued to reside in the structure after her demise. Having visited the town, Swinton made the following observation:

> According to the same author [Balbinus] a spectre, apparition, or ghost, in his time, walked in the neighbourhood of this castle, and even in the castle itself. It was then, if he may be credited, so well known to all the citizens of Neuhaus, as well as the peasants of the adjacent villages, that not the least scruple was entertained amongst them about the reality of its appearance. It was believed to be the ghost of the aforesaid matron, as it appeared in the shape, or form, of a woman, with a bunch of keys hanging at her girdle, and dressed in white; from whence it was called, by the people abovementioned, the White Lady. Several persons of unexceptionable authority affirmed to Balbinus, that they had seen the White Lady, particularly a rector of the Clementine college, who declared, that he once saw her from a window of the castle at noon-day. She then appeared in the market-place all in white, with ribbands about her head, very tall, and with a modest countenance. Her father added, that, when she saw herself discovered by many people, who pointed at her, she grew less gradually, and at last disappeared. Whatever our readers may think of this story, many Bohemians, and some of very good fashion, still believe the reality of this apparition.[33]

Swinton's description of the legend was inspired by the multi-volume work *Miscellanea historica regni Bohemiae* of the Bohemian Jesuit Bohuslav Balbín (the first volume was published in the 1670s). Balbín's version of the legend connects the spirit to a real historical figure, Perchta of Rožmberk, a woman who lived in the fifteenth century. His description became popular, and it still occupies its place in the historical consciousness of the people of the region.[34]

In his travelogue, Swinton introduces the legend of the White Lady to "curious readers" as a story of "several very remarkable particulars".[35] While it is apparent that the author was amazed by the legend, it is difficult to ascertain—as it is in the case of Browne's account of mining—whether Swinton described it as a true wonder or presented it from the perspective of locals who believed in wonders. In order to dispel readers' doubts, Swinton made clear that he based his narrative on information sourced from well-known authors and from testimonies of trustworthy individuals he encountered during his travels. This is a strategy frequently employed by authors who sought to enhance the credibility of their texts. However, it is virtually impossible to ascertain the veracity of the testimonies presented. The trend of addressing the reader is a common feature of eighteenth-century European travel literature.[36]

In order to conclude the discussion on curiosities and rarities in a more practical manner, it would be beneficial to consider a more tangible example. When John Swinton and his English compatriots were leaving Vienna and heading to Prague, a meeting with Mr. Robinson, an English minister,[37] is mentioned in the travelogue:

> Before our departure for Bohemia, he was so good as to suggest to us several hints, for the better regulation of our conduct on the road… He also made us a present of four bottles of Tokay wine, which came out of the Emperor's own cellar. Two of these, being the common sort, appeared like a high-coloured white wine; but the others were of a reddish, or rather a brownish, colour, and an exceeding great rarity even at Vienna itself. All the genuine Tokay wine is said to be preserved for the use of the Emperor only; so that all the wine going by that name in foreign countries, except such as his Imperial Majesty sends as presents to the princes he is in alliance with, cannot be the produce of the mountain and district of Tokay.[38]

Nevertheless, as the author himself points out, the rarity of the Tokay wine was so exceptional that the Germans—Austrians, Bohemians, Silesians and Moravians in particular—were fond of it very much. The first bottle of wine was purloined from English travellers before they crossed the Austrian-Moravian border, which they indicated did not elicit any surprise. They suffered a further loss of the second bottle in an inn in the Bohemian town of Slavonice and resolved to determine the ownership of the remaining bottles the following evening, consuming them in the town of Tábor in southern Bohemia before they reached Prague.[39]

Wonders and the Bohemian cult of saints

In the early modern period, fascination with local marvels, curiosities and supernatural phenomena was an integral part of English travellers' grand tour experience. Yet there is a shortage of documentation relating to descriptions of extraordinary occurrences in early modern English travel accounts, especially with regard to those dealing with the Bohemian Lands.[40] In their descriptions of the region, Browne, Swinton and Pococke sought to capture readers' attention with references to local peculiarities, folklore and traditional legends, but also to baroque manifestations of Catholic piety, an obvious curiosity for Protestant travel writers and their readers.

After the battle of White Mountain in 1620, when the Bohemian King Frederick, a husband of Elizabeth, daughter of the English King James I, the Bohemian Lands became Catholic when, at the battle of White Mountain in 1620, King Frederick was defeated by the Habsburg Emperor Ferdinand II. After this event, English travel writers often comment on the transformation of Bohemia from a multi-confessional and, in their opinion, religiously tolerant country into a mono-confessional Catholic region.[41] The battle of White Mountain and the following change in religion were of particular interest to English readers, because King Frederick was married to Elizabeth Stuart, the daughter of King James I.

English travellers to Bohemia critically reflected on what they considered the zealousness and fanaticism of the Bohemians in matters of faith. This is particularly evident in descriptions of the piety of the locals, which manifested itself in the baroque cult of saints and in a belief in the miraculous power of holy relics. For Protestant observers, who generally did not recognise saints as intercessors and advocated direct appeal to God, the extent of the widespread veneration of Christian saints was surprising. A notable example of such veneration was the cult of St. John of Nepomuk, which was established after the completion of his canonisation in 1729.[42] Following this event, he became the most popular Bohemian saint. Early modern travellers explored places associated with John of Nepomuk (1345–93), especially in Prague, where he had been the vicar-general of Saint Giles Cathedral. Travellers' interest in Nepomuk only increased when we turn from the seventeenth to the eighteenth century.[43] Perhaps the most significant location of interest associated with Nepomuk was the Charles Bridge, from where he was fatally thrown into the river Vltava. Since 1683, this event has been commemorated by a bronze statue of the saint made by the Bohemian baroque sculptor Michael Brokoff. The installation of other sculptures of saints on the bridge at the beginning of the eighteenth century served the purpose of confirming the status of the bridge as a symbol of the new religious identity of the country. Furthermore, the ornaments of the bridge signalled Prague's noteworthy position within the Catholic world and highlighted the importance of the cult of saints to Bohemia and its people.[44]

In his travel journal, Swinton describes how crowds of people prayed on the Charles Bridge, paying special attention to Nepomuk's statue. "There are many curious images or statues of saints upon the bridge over the Moldau", Swinton writes, and they "very well deserve to be seen by every curious traveller, and particularly that of St. John of Nepomuck". In addition to the statue of St John of Nepomuk on the Charles Bridge, "there is a cross of copper or brass deaurated, which people are continually kissing from morning till night, when they offer their prayers to St. Nepomucene, who is esteemed as one of the principal Bohemian Saints".[45] It may be found on the balustrade, placed at the alleged spot where Nepomuk was thrown into the Vltava River. It is closer to the Old Town, between the neighbouring statues of St. Norbert and St. John the Baptist. It is a two-armed bronze cross, the arms of which are crowned with five stars. The cross is set in a marble slab, under which the remains of the saint's body are placed in the railing. The cross is complemented by a decorative wrought-iron grille with a relief of the saint on the surface of the river by the bridge.

On the veneration of Nepomuk, Swinton also observes that "many persons there wear his picture in miniature on their breasts, hanging down like the badge of an order; and most of the women have such a picture, by way of ornament, annexed to their necklaces".[46] Having been drowned at the behest of king Wenceslaus IV, Nepomuk became venerated as a patron saint of bridges with the specific role of protecting people from floods and other dangers. Additionally, he was regarded as an intercessor for women who were unable to conceive.

Swinton was astonished by the attention paid to Nepomuk and other saints. He also mentions that it was possible to buy not only collections of pictures of the statues that decorated the Charles Bridge but also other kinds of souvenirs that often bore the portrait of St. John of Nepomuk.[47] It is necessary to emphasise that Pococke included in his manuscript such a complete set of images of statues and sculptures on the Charles Bridge, and these pictures constitute a significant element of the manuscript. The fact that both he and Swinton visited the place around the same time—in the 1730s—confirms the popularity of such souvenirs in early eighteenth-century Prague. In Pococke's manuscript, there also is a picture of the entire Charles Bridge with its statues of saints. Approximately in the middle of the illustration, the spot from which John of Nepomuk was thrown into the river Vltava is clearly indicated with five stars—the saint's attribution.[48] The dotted line in the picture marks the way from this spot towards the Old Town in Prague, where John of Nepomuk was buried in the now defunct Church of the Holy Cross the Greater.[49] Pococke acknowledges the wondrous that was associated with the saint when he writes that "an excess of devotion they have towards St John Nepomucene, which without doubt was more than ordinary on his first canonization and as they were in danger to loose the excessive of this religion".[50]

In accordance with Swinton, Pococke describes the bridge as the greatest curiosity in Prague, both because of the number of baroque statues and because of the attention given to the sculptures by local inhabitants.[51] In contrast to Swinton and Pococke, those who travelled in the area in the second half of the eighteenth century critically judged the Charles Bridge to be an example of baroque "bad taste". What is more, in their view, the bridge failed to reflect the question of the miraculous powers of saints.[52]

John Pococke's manuscript reveals the author's fascination with the religious practices of the Bohemians. In his observations, Pococke highlights the superstition and habitual overzealousness of Bohemians, indicating, for example, that they had ornated altars in their homes.[53] It is evident from the text that, as a Protestant, he did not comprehend that the exaggerated veneration of saints and other unfamiliar religious acts were manifestations of local piety. It is equally clear, however, that he did understand how common and widespread it was in Bohemia to believe in the miraculous nature of holy images, such as those of John Nepomuk.

The confessional divide between the English travellers and the Bohemians manifests itself occasionally in the travel writers' attempts to safeguard credibility when describing wonders or wondrous phenomena. For example, in *The travels of three Englishmen*, Swinton makes several references to a local man who guided Swinton and his companions around Prague. While the guide had a comprehensive knowledge of the city and conversed with the Englishmen in Latin, it is nevertheless worth mentioning that Swinton frequently cites his statements and thus demonstrates trust in the goodwill and reliability of the guide, who was Catholic. In one instance, Swinton explicitly considers the trustworthiness of the guide and does this with a reference to the Catholicism of the latter: "'tis true, he seemed not a little inclined to bigotry; but, as the abovementioned particulars bear no relation to religion, this will not in the least affect their credibility".[54]

In contrast to the other travel books examined in this study, Swinton makes frequent use of guides in his narration, a practice that is not uncommon in eighteenth-century travel literature. An imaginative guide would accompany the author-traveller on their journeys, becoming an indispensable component of both their observations of unfamiliar lands and their interpretation thereof.[55] According to Swinton's remarks, guides were real persons, either locals or people with good knowledge of the visited region. Their main purpose was to support the credibility of the author's narration when describing the wondrous and extraordinary things of Prague. Swinton points out the abovementioned Catholic guide again when he describes other curious statues (besides the one of St John of Nepomuk), especially when referring to the figures of Martin Luther and his wife:

> Luther appears in armour, and his wife with one of her hands extended towards his privities; which was done in order to ridicule the Lutherans, and perhaps the protestants in general. The citizens of Prague, who are, for the most part, bigotted Roman catholicks, take great care to shew these statues of Luther and his wife to all protestants that come here. This the author of the present account collected from the guide, or, as the Italians call that sort of servant, Cicerone, who took particular care to shew us the abovementioned effigies of Luther and his wife, and desired us to view them attentively, assuring us, that no foreigner, especially if he was a catholic, who knew Prague, would believe that we had seen the capital of Bohemia, if we could not give a description of them.[56]

Statue of Luther and his wife was located on the Old Town Bridge Tower near the Charles Bridge. The story recounted by Swinton was merely a piece of idle chatter that originated among the citizens of Prague in the early eighteenth century. As a matter of fact, the statues in question did not represent either Martin Luther or his wife, Catherine von Bora.[57] The sculpture, described in such vivid detail to the visitors, demonstrates how art could be used to further the interests of the Catholic victory over Protestantism. Swinton himself states that this was an attempt to ridicule Lutherans and potentially Protestantism in general.[58]

Conclusion

The notion of wonder in the context of early modern travel is one with complex ramifications and, in the case of this chapter, involves themes such as confessional differences, local beliefs and authors' credibility. It should also be stressed that wonder and the wondrous went through a significant transformation from the sixteenth to the eighteenth century as the perception of geographical and cultural distance changed. What was perceived as a wonder in the Middle Ages and in the beginning of the early modern period became a mere curiosity for many who travelled in the eighteenth century. Due to the reading audience's constant demand for news and curiosities, travel writers continued to report novelties and extraordinary phenomena. Especially in the era of the Enlightenment, the habit of describing wonders and curiosities raised the question of credibility.

The authors discussed in this chapter employed a number of strategies in order to assert the reliability of their travelogues. John Swinton sought to enhance the objectivity of his text by incorporating citations from established historians (particularly those from antiquity) and first-hand witnesses.

In contrast to Swinton's reliance on external authorities, Edward Browne and John Pococke described extraordinary and wondrous occurrences with a certain objectivity and detachment, leaving it up to the reader to decide whether to believe in the portrayed wonders or not. All three travel journals (including Swinton's work) are written in a style that aspired towards objectivity by mimicking the scientific discourse of the day. At the same time, however, they still brim with passages in which rare and unusual things are perceived and described with curiosity, even wonder. These observations are usually made with references to Bohemian locals who believe in such wonders, while the author's own thoughts remain hidden behind the veil of objectivity.

Notes

1 Browne devoted almost four pages to the description of the cabinet of curiosities in Dresden. See, Browne, *An Account of Several Travels*, 129–132.
2 Nugent, *The Grand Tour*, 243.
3 See, for instance, Swann 2010.
4 The famous example is the *Kunstkammer* of the emperor Rudolf II in Prague. See, Fučíková *et al.* 1997, 166.
5 Jones 2019, 36.
6 De Mello Abdalla 2018, 56–61.
7 On the detailed analysis of the Bohemian Lands as a travel destination in the early modern period, see, Ferencová 2018b.
8 Browne, *An Account of Several Travels*. His travelogue was published on the basis of letters he sent from his travels, especially to his father. These letters are now held in the collections of the British Library, but the notes relating to Bohemia do not contain any additional information relevant to this study. For analysis of Browne's manuscripts and printed books, see Wyatt 2021.
9 For instance, John Harris adapted Browne's text for his *Compleat Collection of Voyages and Travels* (London, 1705). This was published again in the 1764 edition by Thomas Osbourne, who also published John Swinton's travel book. See, Van Strien 2004.
10 This fact is explicitly stated in [Swinton], *The travels of three English gentlemen*, 218: "as a supplement to the ingenious Dr. Browne's Travels in Germany; and will likewise serve to supply various omissions, and illustrate several obscure passages, of other modern relations, with which gentlemen of erudition, who have visited the countries herein mentioned, have entertained the public".
11 [Swinton], *The travels of three English gentlemen*, 218–355.
12 While on John Swinton's life and travels, see Shaw 1999, 295–304.
13 On Pococke's life and travels, see Baigent 2004.
14 The chapter on Bohemia is included in the fifth book called "Observations on Germany, Bohemia, Hungary, Istria, and some parts of Italy". See Pococke, *A Description of the East*, 236–39.
15 Pococke left a travelogue manuscript about every country or region he visited. The original manuscript on Bohemia provides more information than the published text. See Pococke, *The Kingdom of Bohemia*.
16 Stagl 2002, 75.
17 Van Strien 2004; Browne, *An Account of Several Travels*, fol. A.
18 Pococke, *A Description of the East*, 238.
19 The first publication of this kind, preceding the English travel accounts in time, was a modest, anonymous work from 1729, *Denkwürdiges und jetzt lebendes Kayser*

Carols-Baad, published in the Bohemian town Cheb (Eger). This was subsequently reprinted two years later and, due to its popularity, again in Nuremberg in 1734 and 1736. See Bucharovič 2000, 55.
20 Pococke, *A Description of the East*, 238–39.
21 Nugent, *The Grand Tour*, 257–58.
22 Pococke, *The Kingdom of Bohemia*, fol. 100.
23 Pococke, *A Description of the East*, 238–39.
24 Helwig, *Hellwig's map of Silesia.*
25 Browne, *An Account of Several Travels*, 120.
26 On the origins and development of the legend, see Koudelková 2006.
27 Browne, *An Account of Several Travels*, 120.
28 On the history of silver mining in Bohemia, see Hrubý 2024.
29 Browne, *An Account of Several Travels*, 121.
30 Browne, *An Account of Several Travels*. Libuše (Libussa, Libushe) is a legendary eighth-century figure, who is regarded as an ancestor of the Přemyslid dynasty and the Czech nation. According to the legend, she married a ploughman named Přemysl, with whom she founded the dynasty and, following her own prophecy, the city of Prague.
31 Mączak 1995, 287–91.
32 [Swinton], *The travels of three English gentlemen*, 352–53.
33 [Swinton], *The travels of three English gentlemen*, 292.
34 Balbín, *Miscellanea historica*, 294–96.
35 [Swinton], *The travels of three English gentlemen*, 292.
36 On the question of credibility in travel writing, see Schulz-Forberg (ed.) 2005, 30; Carey 2019, 524–47; and most recently, Van Groesen and Müller (eds.) 2024, particularly the introduction.
37 Thomas Robinson was the English ambassador to Vienna in the 1730s and 1740s.
38 Swinton, *The travels of three English gentlemen*, 265. Tokay wine is traditionally produced in north-eastern Hungary and south-eastern Slovakia.
39 [Swinton], *The travels of three English gentlemen*, 265, 283, 291, 293.
40 Mączak 1995, 287–91.
41 For a more thorough analysis of the early modern British perception of the Bohemian religious transformation, see Ferencová 2021; Ferencová 2018a. On the situation in Bohemia, see, for instance, Louthan 2009, 1–15.
42 For an overview of the proceedings concerning the saint, see Vlnas 1993, 98–127.
43 Louthan 2010, 177–78.
44 Louthan 2010, 178–79.
45 [Swinton], The travels of three English gentlemen, 296–97.
46 [Swinton], The travels of three English gentlemen, 297.
47 [Swinton], The travels of three English gentlemen.
48 According to the legend, the body of John of Nepomuk was discovered in the river Vltava on 17 April 1393, allegedly surrounded by five lights. These were instrumental in facilitating the discovery of the body. Consequently, the halo of John of Nepomuk is adorned with five stars, which constitute his most celebrated attribute.
49 The picture is folded and inserted like a separate sheet in the manuscript; Pococke, *The Kingdom of Bohemia*, fol. 49.
50 Pococke, *The Kingdom of Bohemia*, fol. 118r.
51 Pococke, *The Kingdom of Bohemia*, fol. 117.
52 For instance: Wraxall, *Travel diary*, 17; Lynch Piozzi, *Observations and reflections*, 316. On the perception of the matter by foreign travellers in general, see Blažek, Mikulec and Pokorný 2023.
53 Pococke, *The Kingdom of Bohemia*, fol. 118r.
54 [Swinton], *The travels of three English gentlemen*, 309.
55 See Carey 2019.

56 [Swinton], *The travels of three English gentlemen*, 296.
57 See Svátek 2002, 111.
58 See Louthan 2008, 72–75.

Bibliography

Primary sources

Anonymous, *Denkwürdiges und jetzt lebendes Kayser Carols-Baad*. Eger: J. A. Orwansky, 1729.

Balbín, Bohuslav, *Miscellanea historica regni Bohemiae* (*Krásy a bohatství české země*), edited by Helena Businská. Prague: Panorama, 1986.

Browne, Edward. *An Account of Several Travels Through a great part of Germany*. London: Printed for Benj. Tooke, 1677.

Harris, John. *Navigantium atque itinerantium bibliotheca, or, A compleat collection of voyages and travels*. London: Printed for Thomas Bennet, 1705.

Helwig, Martin. *Hellwig's map of Silesia*, 1561. Geoportal ČÚZK. Accessed 29 May, 2024. https://geoportal.cuzk.cz/(S(apsuvvlvxime3liw1rcapx4u))/Default.aspx?lng=EN&mode=TextMeta&side=mapy_tiskKHM&metadataID=CZ-CUZK-HELLWIG-T.

Lynch Piozzi, Hester. *Observations and reflections made in the course of a journey through France, Italy, and Germany*, vol. II. London: Printed for A. Strahan; and T. Cadell in the Strand, 1789.

Nugent, Thomas. *The Grand Tour. Containing an Exact Description of most of the Cities, Towns, and Remarkable Places of Europe*, vol. II. London: Printed for S. Birt, 1749.

Pococke, Richard. (n.d.), *The Kingdom of Bohemia, Bohemia: Travels in, by Rev. R. Pococke: 18th cent. (1736–1737), vol. XIII (unpublished manuscript)*. Add MS 22990, British Library, London.

Pococke, Richard. *A Description of the East and Some Other Countries*, vol. II, part II. London: W. Bowyer, 1745.

[Swinton, John]. The travels of three English gentlemen, from Venice to Hamburgh, being the Grand Tour of Germany, in the year 1734. In *The Harleian Miscellany: Or, A Collection of Scarce, Curious, and Entertaining Pamphlets and Tracts*, vol. XI., edited by William Oldys, John Malham and Robert Dutton, 218–35. London: Printed for Robert Dutton, 1810.

Wraxall, Nathaniel William. *Travel diary, 1777–1779. Osborn Collection c23, Beinecke Rare Book and Manuscript Library*, New Haven, USA.

Literature

Baigent, Elizabeth. "*Pococke, Richard (1704–1765), traveller and Church of Ireland bishop of Ossory, of Elphin, and of Meath*". Oxford Dictionary of National Biography, 23 Sep. 2004. Accessed 9 May 2024. https://www.oxforddnb.com/view/10.1093/ref:odnb/9780198614128.001.0001/odnb-9780198614128-e-22432.

Blažek, Petr and Mikulec, Jiří and Pokorný, Vojtěch. *Musica Navalis: Dějiny slavností a kultu sv. Jana Nepomuckého [Musica Navalis: The History of Festivities and the Cult of Saint John of Nepomuk]*. Prague: Svatojánský spolek, 2023.

Bucharovič, Stanislav. *Karlovy Vary a jejich vlastivědné písemnictví: průvodce světem carlovarensií [Karlovy Vary and its historical sources: a guide to the world of Carlsbad].* Karlovy Vary: Státní okresní archiv Karlovy Vary, 2000.

Carey, Daniel. "The Problem of Credibility in Early Modern Travel". *Renaissance Studies* 33, no. 4 (2019): 524–47.

De Mello Abdalla, Frederico Tavares. "Travel Instructions in the second half of the 18th century: the making of "how-toobserve-and-collect" texts in Portugal". In *The scientific dialogue linking America, Asia and Europe between the 12th and the 20th Century. Theories and techniques travelling in space and time*, edited by Fabio D'Angelo, 56–67. Naples: Associazione culturale Viaggiatori, 2018.

Ferencová, Hana. "'Who remained, turned their Religion': Bohemian Lands in the Travel Journal of John Swinton". *Acta Comeniana* 32 (2018a): 93–113.

Ferencová, Hana. "Obraz konverze českých zemí v britských cestopisech" [The image of conversion of the Bohemian Lands in British travel journals]. In *Vytváření konvertity. Jazyková a vizuální prezentace konverze v raném novověku* [Verbal and Visual Representation of Conversion in the Early Modern Era], edited by Radmila Prchal Pavlíčková, Iveta Coufalová and Hana Ferencová, 161–185. Prague: NLN, 2021.

Ferencová, Hana. *Čechy a Morava očima anglických cestovatelů 1570–1800 [Bohemia and Moravia in the Eyes of Early Modern English Travellers 1570–1800].* Olomouc: Vydavatelství Univerzity Palackého, 2018b.

Fučíková, Eliška et al. (ed). *Rudolf II and Prague: The Court and the City.* London: Thames & Hudson, 1997.

Hrubý, Petr. *Silver Mining in the Kingdom of Bohemia (13th-14th Centuries).* Leiden: Brill, 2024.

Jones, David. *From Wonder to Curiosity in Early Modern Travel Writing.* PhD diss. University of Liverpool, 2019.

Koudelková, Eva. *Krakonoš v literatuře: kapitoly k literárnímu ztvárnění látky o Krakonošovi [Krakonoš in literature: chapters on the literary representation of Krakonoš].* Liberec: Nakladatelství Bor, 2006.

Louthan, Howard. "Religious Art and the Formation of a Catholic Identity in Baroque Prague". In *Embodiments of Power: Building Baroque Cities in Europe*, edited by Gary B. Cohen and Franz A. J. Szabo, 53–79. Oxford – New York: Berghahn Books, 2008.

Louthan, Howard. "Tongues, Toes, and Bones: Remembering Saints in Early Modern Bohemia". *Past and Present* 206 (2010): 167–83.

Louthan, Howard. *Converting Bohemia. Force and persuasion in the Catholic Reformation.* Cambridge: Cambridge University Press, 2009.

Mączak, Antoni. *Travel in Early Modern Europe.* Cambridge: Polity Press, 1995.

Schulz-Forberg, Hagen (ed.). *Unravelling civilisation: European travel and travel writing.* Brussels: Peter Lang, 2005.

Shaw, Trevor. "John Swinton, F.R.S., Identified as the Author of a 1734 Travel Journal", *Notes and Records of the Royal Society of London* 53, no. 3 (1999): 295–304.

Stagl, Justin. *Eine Geschichte der Neugier: die Kunst des Reisens 1550–1800.* Wien: Böhlau, 2002.

Svátek, Josef. *Pražské pověsti a legendy [The Tales and Legends of Prague].* Prague: Paseka, 2002.

Swann, Marjorie. *Curiosities and Texts: The Culture of Collecting in Early Modern England*. Philadelphia: University of Pennsylvania Press, 2010.

Van Groesen Michiel, and Johannes Müller (eds.). *Far from the Truth: Distance, Information, and Credibility in the Early Modern World*. New York: Routledge, 2024.

Strien Strien, Van., "*Browne, Edward (1644–1708), physician and traveller*". Oxford Dictionary of National Biography, 23 Sep. 2004. Accessed 9 May 2024. https://www.oxforddnb.com/view/10.1093/ref:odnb/9780198614128.001.0001/odnb-9780198614128-e-3670.

Vlnas, Vít. *Jan Nepomucký, česká legenda [Saint John of Nepomuk. Czech legend]*. Prague: Mladá fronta 1993.

Wyatt, Anna. "*'Wide Excursions': Dr Edward Browne (1644-1708) and the Writing of Travel in Restoration England*". PhD diss. University of East Anglia, 2021.

9

REDUCING AND OTHERING WONDER

The Work of John Green and Abbé Prévost

Antoine Eche

As the eighteenth century developed, so did the taste for travel in European reading circles, travel accounts being recorded in France as the most preferred genre as early as the late 1690s. News and information about distant lands would feed the reading population's growing appetite for, broadly speaking, things out of the ordinary and also sustain on the travellers' side a more fundamental attitude pertaining to the animal kingdom as evidenced by psychologist Daniel Berlyne, i.e. curiosity and its deep connection with locomotion.[1] As we know, curiosity in Europe was not, however, considered a positive attitude since at least Saint Augustine while wonder—also attached to things deemed out of the ordinary—was praised and made explicitly available in books of mirabilia and through an important imagery of marvellous and extraordinary creatures and phenomena associated with the margins of the known world, the faraway. Distance from the more familiar centre established the condition for the astonishing, the surprising. Both concepts were, however, used for the construction of knowledge and their respective "behaviour" shared many similar traits.[2] As Lorraine Daston and Katharine Park have pointed out, "wonder and curiosity interlocked" in the seventeenth century until wonder and wonders began losing their appeal in the eighteenth century.[3] It seems, therefore, unavoidable when reflecting on manifestations of wonder and representations of wonders in sixteenth- to eighteenth-century travel literature to articulate these two concepts together. Wonder and curiosity as such need to be studied not as a stable feature of the travellers' *outillage mental* (mental tool set) but as shifting attitudes/concepts inscribed in the transformations of the period's Western epistemology.

DOI: 10.4324/9781003506928-9

This chapter will, therefore, examine how the inflection of the attitude towards wonder and wonders in the first half of the eighteenth century can be perceived and evaluated within this context through a developing form of travel literature, i.e. collections of travels and voyages. Although not entirely novel in the travel literature landscape of the period, collections clearly benefited from the increasing taste for the genre but also from the developing publishing market and the proliferation of what Roger Chartier calls "*bibliothèques sans murs*" (libraries without walls).[4] Works aiming to accumulate all knowledge on specific subjects, such as encyclopaedias, dictionaries, collections of all sorts and universal histories, surged between the 1680s and the 1750s, and travels accounted for a large part of this production, almost always including numerous illustrations. These collections offered to the eyes of the curious readers the spectacle of the extra-European world. We will focus specifically on what was probably the most well-known collection of the period: the *Histoire générale des voyages*, by Abbé Prévost. Published from 1746 until 1759, it is the translation and continuation of John Green's *New General Collection of Voyages and Travels* (1745–1747). The choice of a translated collection offers a different perspective on the way that the contents could be apprehended in a critical and reflexive fashion by contemporary readers of such collections. The translator, through his work, develops a criticism of the source text, whether consciously or not, including its contents and its sociological and ideological framework. As such, the translator's lens provides insights into epistemological trends. As we shall see, the investigation on the inflection of wonder and wonders will benefit from this particular perspective, but not from it alone. While we will examine how these two compilers, following the epistemological trend of the period, reduced and critiqued the manifestations of wonder and wonders contained in some of the accounts used in the compilation, we will also show how this attitude may also contribute to the othering of wonder and wonders in order to establish and/or reinforce what Walter Mignolo calls the "external colonial difference", understood as the construction of the alleged ontological inferiority and the epistemic disability of non-European populations during the colonial period.[5] It is well-known that extra-European travel writing at this period played a significant part in the European colonial agenda. As manifestations of a collective sensibility, wonder and wonders could not be immune to this ideology as "the experience of the domination of the centre over a periphery", and they contributed to the making of modern subjectivity, permeating the whole social fabric.[6] Before addressing this issue, we will need to introduce the particulars of Abbé Prévost's collection as well as clarify the presence and functions of wonder in travel writing.

From Green's *Collection* to Prévost's *Histoire*

The *Histoire generale des voyages*, although consistently associated with Abbé Prévost, is not originally Prévost's work but that of an obscure cartographer,

John Green, and was published over three years in London as *A New General Collection of Voyages and Travels.*[7] The translation was ordered by the Chancelier d'Aguesseau for the Minister of the Navy, the Count of Maurepas.[8] At the time, Abbé Prévost—who is now mostly remembered for *Manon Lescaut* published in 1731—was a public figure and a prolific man of letters: a novelist, journalist and translator. He had spent some time in England (1728–1730) where he learnt the language, wrote and almost got jailed for forgery and the falsification of documents. He moved to Holland, returned to England and also resided in France. He led a tumultuous life, went bankrupt, became the chaplain of the Count of Conti, went bankrupt a second time and had a *lettre de cachet* issued against him. When he returned to France in 1742, he led a quieter life. The translation of the English collection kept him busy for 13 years and provided him with financial stability. Prévost was not new to translation, having already transposed historical works written in Latin and English into French, as well as literary miscellanies from English to French. These activities gave him a solid reputation as a translator in France.[9]

The French version is not exactly the translation of the English original but can be seen as its continuation and conclusion as well. Indeed, the English editor, John Green, as a minor cartographer with a troubled past, failed to raise sufficient interest in his project and had to abandon all hopes of being able to complete it because of a lack of financial security and possible political entanglement.[10] When the English version was curtailed, Prévost decided to take up the mantle while reducing the scope of Green's ambitious project. This explains the discrepancy in terms of volumes in each collection: the English version consists of four quarto volumes as opposed to the 15 quarto volumes of its French counterpart.[11]

We are able to distinguish three different editorial phases for the French version, which correspond to three distinct periods and volumes: 1746–1750 (vols. 1–8); 1750–1753 (vols. 8–11); 1754–1759 (vols. 12–15). We also consider two translation periods, the first one ranging from 1746 to 1750 (vol. 1–8) and the second one from 1750 to 1759 (vol. 8–15).[12] During the first period, Prévost translated the material already collected by Green while later he also had to collect it first and sometimes translate it if he was not using French or already French-translated texts. These two moments correspond to a difference in terms of the composition of the global collection. In the final period, Prévost, who had been complaining against the outline designed by Green since the first volume, modified the original structure, however, noting that he could not completely emancipate himself from the English design in order to produce the collection that he would have liked. In terms of geographical coverage, the English collection encompasses Africa (vols. I–VI) and China, Korea and Tibet (vols. VI–VII). To this, which amounts to the first seven volumes of the French translation plus another 60 pages in volume VIII, the French collection adds the East Indies (vols. VIII–XI), Southern and Antarctic territories and various areas (vol. XI), the Americas (vols. XII–XV) and the travels to the North-East and North-West as well as the French West Indies (vol. XV).

In spite of these differences, both the original and the translation/continuation relied on a body of selected travel accounts and geographical treatises. For both Green and Prévost, the *raison d'être* of the collection was "the improvement of geography, navigation and natural history, principally in view, by bringing together a sufficient quantity of materials, for making accurate maps, charts and descriptions of foreign countries" in order to produce "a system of modern geography and history, as well as a body of voyages and travels".[13]

Clearly, the English version was already of note in the context of geographical literature in England at the time. It offered unabridged versions of known texts, additions of recent travel accounts, fresh cartographical knowledge and a critical approach that would distinguish between the adventures of the travellers and their observations, thus providing the reader with a summary of complementary and also contrasting representations of the faraway. A brief example should suffice here to show the thoughtfulness involved in the design of this collection:

> Although our design is much more extensive than that of any collection hitherto published, yet we propose to execute it in less room than any of the former. To effect this, we have deviated from the common method of collecting, and instead of giving each author entire in the order he was published, we separate his journal and adventures from his remarks on countries: the first we give by itself; the latter we incorporate with the remarks of other travellers to the same parts. The adventures of travellers are generally very tedious; often trifling, and therefore admit of large retrenchments; and as several travellers visiting the same parts must necessarily repeat the same things, it is certain, that by this way of collecting them, a vast deal of superfluous matter will be expunged.[14]

What must be mentioned, to Green's credit, is that he had already envisioned the original idea for the conception of this collection 30 years before its publication. Indeed, Green is thought to be the author of a treatise of cartography entitled *The Construction of Maps and Globes* (London 1717), in which the author expresses the wish to one day see a collection that would follow a list of certain principles. These principles are the ones exposed and put to use in Green's collection.[15] The table of contents reflects this system as it usually displays two sets of entries, the first ones identifying the narratives with explicit references to the form or to the idea of motion itself—for instance "account", "letter" or "voyage" and the second ones referring to ethnographic content, descriptions of the fauna and flora, etc. This design was also complemented with a somewhat thorough cross-referencing system that Green developed in the summaries of the travellers' descriptions, contrasting different sources and very often adding comments of his own in the footnotes. Prévost followed suit, except for the notes that he removed when he found Green's remarks too anti-Catholic or not appropriate in terms of tone or because of diverging views.

This critical approach, even if not perfect, and the freshness of this collection are probably the reasons why it caught the attention of the French. Another reason would be the fact that this translation filled a gap in the context of French collections of voyages and travels, as there were no recent examples that could be used in the context of France's colonial expansion. When it was published, the translated collection received very good reviews, which praised the initiative and the method used by the English author and the translator.[16] The *Histoire générale des voyages* quickly became a success (a pirate edition was published as early as 1747 in Holland) and was then expanded even after Prévost completed the original final volume of the collection. An abridged version of 21 volumes was also published by Jean-François de La Harpe in 1780. Prévost's collection was used by men of letters, philosophers (Rousseau, Voltaire, the Encyclopedists, Buffon, etc.) and artists looking for a variety of information about distant places and their populations. Rousseau's definition of the savage man in the second discourse (1755) relies, among other things, on an extensive quote taken from the French translation. Abbé Marc-Antoine Laugier, mostly known for his *Essai sur l'architecture* (1753), also insisted, in other publications, on the quality of the prints found in the *Histoire générale des voyages* and encouraged artists to peruse this publication in order to enhance their visual rendition of the faraway in their work.[17] Both the text and the images were acclaimed. Notwithstanding, some critiques were addressed to Prévost, notably by the Dutch editor of the pirate edition, who accused Prévost of transforming the original and also by authors who were later in possession of more accurate information about regions covered in the first part of the collection.[18] Nonetheless, the success of the French version continued until the French found themselves occupied with domestic affairs towards the end of the 1780s.

Travel Writing and the Topos of Wonder

When wonder is discussed in connection with travel writing, it is often defined as a "stock trope [transposing] the emotional and intellectual response that occurs when a traveller is confronted with something new that temporarily defies understanding, and that cannot easily be assimilated into the conceptual grid by which the traveller usually organises his or her experience".[19] As Francois Hartog has demonstrated, this topos was already part of Herodotus' rhetorical tools; the idea of the extraordinary was rendered through a double structure that articulated quantitative and qualitative terms using the lexicon of measure and evaluation.[20] The marvellous became a staple of the Western world and solidly established itself in the Western imaginary during the Middle

Ages and the Renaissance, especially in its association with travel.[21] Although there is no explicit poetics of travel writing, wonder is understood to be a stylistic component of travel accounts to render travellers' reactions into novelties.[22] Novelty necessarily had to be appraised through an already fairly developed conceptual grid constituted by the travellers' socialization, including their education, their knowledge, their experience, their beliefs, their prejudices and their feelings.[23] It is with this material that they were able to proceed to what Jean-Didier Urbain calls the "*mise en intrigue du monde*" (the emplotment of the world) and naturally, of distant places.[24] To paraphrase Urbain, before going somewhere, the traveller necessarily had to imagine this other place.[25] Because it contributed to the making of a new and different place, this individual and collective construct cannot be solely understood as a means of protection for the traveller against the impact of the novelty.[26] As already a part of the sensibilities involved in the emplotment of the world, wonder was therefore included in the travellers' apprehension of distant places before they would even arrive at the desired destination. Between the sixteenth and eighteenth centuries, the writing of the experience of travel formalized this emplotment up to a point that might remain implicit for the reason that the act of writing was conditioned by what Réal Ouellet calls the "*pacte viatique*" through which travellers would endorse the colonial project and also legitimize a way of writing.[27] This pact shaped not only the contents of travel accounts but also what can be labelled generic conventions of the tradition: in the absence of an explicit poetics of travel writing, this mixed genre constituted itself with precepts taken from various referential genres (history, correspondence) besides formalized humanistic travel practices and the development of the sciences in the seventeenth century.[28]

The iteration of wonder in writing as a topos takes place within other descriptive and narrative tropes such as departure scenes, mentions of preparations, scenes of encounter, anecdotes, etc.[29] Wonder, therefore, appears in the text as the result of a double mediation: first through the emplotment of the world and then through the necessity for travellers to have their accounts identified from a generic point of view as genuine travel accounts, including all the expected components. Such expectations were not necessarily immutable, as they were based on society's epistemic modulations and trends. Wonder in medieval travel writing is not identical to wonder in seventeenth- and eighteenth-century travel accounts. The notion had to conform to the Foucauldian "*regime de veridiction*". In other words, it had to be acceptable according to the paradigm of truth and falsehood of the day. This is why Samuel Sprat in the *History of the Royal Society* criticizes the marvellous aspect of some accounts used by scientists: while once deemed acceptable, such a conception of wonder could no longer be tolerable for Sprat's contemporaries.[30]

At the same period, a semantic shift has been observed regarding curiosity. While previously considered a human desire, curiosity now started to designate objects of knowledge as well.[31] In so doing, curiosities became wonders.[32] Such objects were qualified as curious, astounding, surprising, rare, new or extraordinary, overlapping the definitions of objects of wonder. This development is epitomized by the cabinets of curiosities, so popular at the end of the seventeenth century and characterized by their classification system, which distinguished between objects of nature, human productions, exotic objects and antiquities.[33] This shift also happened in travel writing.[34] However, in contrast to the cabinets of curiosities, in travel writing, curious and astonishing objects were not totally extracted from their original cultural space and juxtaposed to other curiosities, but they were included in the exotic space built by the narrative. Once measured, described, ideally understood and finally digested by the Western observer, this new, rare and curious object would cease to be what it was. Following the increased taste for travel literature and the ensuing growth of its market, curious and astonishing objects became legitimate categories of travel accounts as long as they did not undermine norms of acceptability. Many titles of relations at the end of the seventeenth century contain the word "curious".[35] In the case of Prévost's collection, the term did not appear on the title page but in the advertisement for the launch of the publication. In a more and more competitive market, curiosities became a label that was thought to encourage sales and respond to the readers' appetite for exotic locales. There is a clear connection between the consumption of distant places and peoples by readers and the appropriation of these places and peoples by travellers and the powers that mandated them and benefited from those travels. To the desire of naming, categorizing and exploiting alterity corresponded the desire of consuming it by proxy. In 1694, the *Dictionnaire de l'Académie* offered the following definition for curiosity: "passion, desire, eagerness to see, learn, possess rare, singular, new things".[36] The notion of ownership included in this definition is not fortuitous. It shows that curiosity/curiosities and wonder/wonders were all understood in relation to possession. As the development of travel literature and its market was coextensive to the progress of colonisation and coloniality, it is striking to note that the congruence of these histories (of travel writing, curiosity, wonder, colonisation, book) leads us to analyse the scenography of wonder in a colonial sense, a colonial dynamic involving the traveller as well as the reader. Once organized and formatted, wonder became foreshadowed and then instrumentalized. It enhanced distant places for personal gain (social advancement of the traveller) and for colonial purposes (accumulation of wealth through extraction; production and development of geographical knowledge in order to gain power). The instrumentalized wonder/alterity relation reinforced what South-American scholars call the external colonial difference in the sense that the Other and their locale were kept at a distance both spatial and temporal. Wonder was used to attest and legitimize the "reality" of the colonial difference.

Othering Wonder

We will now argue that there is a way to read wonder and wonders in this collection, which manifests itself in the peritext and the epitext. A comparison of the titles of both versions suggests how the notions of wonder and wonders need to be understood when reading this collection. Prévost does not change the title inasmuch as he puts to the fore certain aspects already mentioned by Green, most notably the historical dimension of the collection. He also expands on the characterization of the contents (the inventory): originally presented as "remarkable", it becomes "most remarkable, most useful and best verified".[37] The prospectus that preceded the publication of the French version contained the same wording as the one found on the title page. An abridged version of this prospectus, however, is used in an advertisement for the books available from the French bookseller, Didot. The description of the collection offers a slightly different wording as it replaced "*remarquable*" with "*curieux*". As we saw above, the use of this adjective in French to qualify a travel book was a commonplace and highlighted the alleged novelty and rareness of the inventory's items, as well as their extraordinary nature, but also a marketing tool.[38] While the wonder and wonders to be found in this collection appear to be somewhat traditional, they are subjected to the overarching principle of historical writing under which travel writing operates. And it is through this principle, as we shall see, that both Green and Prévost reduce wonder and wonders. But before going any further, it must be stressed that wonder and wonders will not be eliminated from the collection. Actually, one could say that they are ineradicable because, as themes, they are ingrained components of travel accounts. In order to be identified as a travel account, the text needs to display, among other things, the theme or trope of wonder. The necessity for wonder and the imperative of referential writing create a tension that is sometimes explicitly expressed by the traveller. We find a very telling example of this situation in Charles Marie de La Condamine's journal of his scientific journey to Peru. In a scene taking place as he is approaching the city of Quito, La Condamine employs an obvious rhetorical trick, expressing the tension between the transcription of the emotional reaction of the traveller, the generic conventions of travel writing and the scientific and rational attitude that also needed to be conveyed to the reader.[39] While wonder is described as a mixed feeling of aston ishment and admiration, it is succeeded by the progressive development of pleasure as he gazes down at a dale interspersed with brooks and cultivated fields, woods, plains and hamlets that remind him of the French countryside. In the original, La Condamine abruptly concludes his evocation with the sentence "*je me suis laisse entrainé au souvenir de la première impression que je reçus alors: j'oublie qu'il n'est ici question que de travaux académiques*" (I let myself carried away with the memory of the first impression that I then felt: I forget that this is not the place for anything other than academic works).[40] Prévost paraphrases the sentence while making the traveller's feelings explicit: "*L'Académicien se reproche ici de se livrer trop à l'ancienne impression d'un si beau spectacle*" (The Academician reproaches himself for his infatuation with the old impression of such a beautiful spectacle).[41]

La Condamine's awareness of his transgression reveals how travellers were bound by traditional images, the readers' horizon of expectations and generic constraints. In his case, he chose to tone down wonder in opposing it to the "more serious" objective of his account. It is precisely this attitude that John Green and Abbé Prévost will develop in the collection while still leaving room for wonder and, to a certain extent, wonders. Needless to say, this rational and scientific attitude also needs to conform to the "*régime de véridiction*", as demonstrated by the reception of James Bruce's account of his travels to Abysssinia. As Charles Batten put it, "Bruce's problem resulted from his failure to follow a well-established rule for travelers: 'It is not sufficient to write things true, but they must likewise seem probable, to gain belief'".[42]

The most basic technique used by Green and then Prévost to truncate wonders was cross-referencing. As mentioned previously, Green's approach was based on the separation of the adventure and the inventory and the combination of the latter with other inventories, thus allowing for contrasting views on similar or identical items. What a traveller would assert about something would be confronted with what other travellers wrote about it. For instance, a brief paragraph concerning the description of the Wolof people in today's Senegal involves descriptions by four travellers. Green doesn't seem to take sides and lets the reader form their own opinion. Everything is clearly referenced in numerous footnotes, in which he occasionally comments on what is asserted by the travellers.[43] The English compiler is particularly prompt at commenting on religious practices and beliefs. For instance, in the abstract of captain Richard Jobson who travelled up the river Gambra in 1620–1621, Green severely judges the captain for ending up believing in the existence of a devil called Ho-rey, despite having first-hand evidence that Ho-rey was in fact a trick played by the local priests to maintain some kind of social control over the boys. But before judging him, Green compares the story told by Jobson with a story from another traveller's account in order to debunk its wondrous aspect.[44]

Prévost's translation gives a different perspective. Generally speaking, it must be said that he simply eliminated most of the footnotes appearing in Green's version, and such is the case here. However, he adds a sentence at the end of the main text where he describes Jobson's story as "surprising" because the captain had been described as "a sensible man" who was able to suspect fraud, and yet the story revealed him lacking basic skills of discernment.[45] Even if Prévost is less assertive than Green, he nonetheless drives the point home. And in one instance in particular, he questions a wonder that Green seems to have missed. The wonder in question is located in one of the texts retained by Green for the section devoted to the Canary Islands, a locale which had for a long time fascinated and puzzled navigators sailing along the Western coast of Africa. What first caught the attention of voyagers was the famous volcanic mountain, Mount Teide. In order to be thorough in his description, Green uses three different texts, written in three different periods: *Observations of the Right Worshipful Sir Edmund Scory, Knight, of the Pike of Tenerife* (Purchas's *Pilgrimage*, 1626); *A Relation of the Pico Tenerife, received from some considerable merchants* (T. Sprat, *History of the Royal Society*, 1667); and *Account of a journey from the port of Oratava in Tenerife to the top of the pike in that island* (J. Edens, *Philosophical transactions*, 29, 1715). The mountain clearly was an object of curiosity and wonder for travellers, especially for Edmund Scory and J. Edens. According to Scory, "the pike strikes one with amazement, both near and at a distance".[46] In similar vein, Edens writes that "the author made his business to admire the strangeness and vastness of that great body, which (says he) is very wonderful: insomuch, that it is impossible to express to one, who never saw it, in what manner that confused heap of rubbish lies; for it may very well be titled one of the greatest wonders in the world".[47]

The wonder that Prévost questions is first mentioned in Scory's account and relates to the ethnographic description of the local population, the Guanches, completely annihilated during the Spanish colonial occupation. Scory's description of the Guanches draws at least from Jean Bethencourt's *Le Canarien* (ca. 1410–1420), a late medieval document that provides a portrait of the population.[48] In addition, he was able to make use of his own experience since he stayed in the region towards the end of the sixteenth and the beginning of the seventeenth century.[49] His description focuses on one aspect in particular, the extraordinary size of the Guanches:

> They were handsome in stature, well-formed, and of a good complexion. There were some giants of an incredible bigness. The skull of one is remaining, in which there are eighty teeth, and his body, (which was found in the sepulchre of the Kings of Guymur, being of that race) measured fifteen feet.[50]

Curiously, Green does not comment on this whereas in the same page he adds a footnote to criticize Europeans like Bethencourt—who was one of the first Europeans to describe the Guanches—because they would systematically consider "people of distant countries" as atheists (the religious practices of such people were denied the status of religious practice). Green then adds that "there is not, perhaps, a nation in the world, however barbarous or inconsiderable, that does not believe in a supreme being, a principle, which necessarily takes place where there are the least glimmerings of reason".[51] On the other hand, Prévost, who promptly removed the attack on religion from the original footnote while keeping the critique of such travellers, questions the truth of Scory's account of the Guanches. In a footnote, he states that "*ce qui regarde ici le géant de quinze pieds parait une exagération*".[52] It is not clear if Green's blindness was a mere oversight, but what is certain is that one might be too quick at judging from this that Canarian giants were not part of Prévost's "*régime de véridiction*". Indeed, the size of the Guanches is mentioned again in the description of their funeral rites, in particular the mummification of the dead, which, according to Green, is a "curiosity not much less surprising than the Pike itself".[53] Now I should add that the description of the caves and the mummies comes from the second text mentioned earlier, which happens to contain the account of merchants and the account of a physician who resided for some 20 years on the island. However, only one narrative voice is used throughout, the merchant's, while the second account is told in reported speech, implying that it is the transcription of a conversation between the merchants and the physician. The evocation of the mummies comes from the second account in an anecdote where the physician narrates how he discovered a cave when he was hunting with local people. In the translation of this story, Prévost modified the description of a mummy and actually reactivated the idea of wonder about the size of the Guanches. The scene opens when one of the locals enters a cave and is described as being suddenly surprised:

[...] he cried out. It was at the sight of these corps, very tall and large, lying with the head on a great stone, the feet supported with a little wall of stone, the body itself resting on a bed of wood.[54]	*Il y entra; mais sa frayeur se fit connoître aussi-tôt par ses cris. Il y avait apperçu un cadavre d'une grandeur extraordinaire, dont la tête reposait sur une pierre, les pieds sur une autre; et le corps sur un lit de bois.*[55]

His translation of "tall and large" by "*grandeur extraordinaire*" goes against his previous reaction to the first mention of giants on the island. His change of mind was possibly motivated not just by the very presence of mummies in these caves but also by the fact that they were presented as being in a perfect state of conservation. We should note here that the main object of wonder was not quite the same for both compilers. What caught the attention of Green was that the goat skins used to wrap the bodies were perfectly seamed together, whereas for Prévost, it was their state of conservation:

"The corps are sewed up in goat skins, with thongs of the same, in a very curious manner; particularly as to the seams, which are incomparably even and exact. The skins are made very tight to fit the corps, which, for the most part, are entire".[56]	"*Les corps y sont cousus dans des peaux de chevres, avec des courroies de la même matiere, et les coutures si égales et si unies qu'on en peut trop admirer l'art. chaque enveloppe est exactement proportionnée à la grandeur du corps. Mais ce qui cause beaucoup d'admiration, c'est que tous les corps y sont presqu'entiers*".[57]

That said, we need to bear in mind that in both cases their wonder and admiration were deeply rooted in the distance that they perceived between the level of skills needed to achieve this craft and the level at which they implicitly placed this population in comparison to their own standpoint. Such art was not expected in these locales, and that is why it became an object of wonder.[58]

These two brief examples show how Green and Prévost truncated the wonders reported by travellers, thus excluding them from what was deemed acceptable in terms of factual truth. While, in the first example, Prévost's comment only focuses on the traveller, Green's remark about Captain Jobson goes further and encompasses a large number of individuals connected by their alleged lack of reason. We can see here how Green's position on religion allowed him to group non-European as well as European peoples on this issue. This said, it does not mean for him that all people are equal: the Catholics are one of his favourite targets while on many occasions, he endorses stereotypical views reinforcing the alleged inferiority of extra-European populations, i.e. non-whites. What remains, also in Prévost's translation, is the constant Eurocentric view according to which non-European knowledge and beliefs are considered not as advanced as those of Europeans, as implied in the second example.

Conclusion

Determined to establish a history of voyages and travels, both men position themselves as people of authority on the subject, and their selection of facts and information needs to appear beyond reproach, perfectly legitimate and validated. Not only do they reproduce Eurocentric, racialized views, but they also legitimize and disseminate them on a broad scale. And the question of wonder and wonders represented in the collection is of prime importance here as its sanctioned perception and understanding reinforce the external colonial difference, as these populations, under a European lens, are seen as "ontologically inferior and epistemically disabled". The necessary representation of wonder and wonders in the collection and its translation ultimately reveals the articulation between tropological and rhetorical expectations and the colonial mindset of the day, even if neither of the authors was aware of it as such. Legitimizing the alleged superiority of a "civilised" centre opposed to a peripheral and inferior "barbarous" alterity, wonder does not seem to be working so much as "a goad to inquiry" than as a tool for validating bias. In line with this observation, wonders need to be seen as "objects of [colonial] investigation".[59]

Notes

1 Stagl 1995, 1–7.
2 Kenny 1998, 14.
3 Daston and Park 1998, 15, 18–19.
4 Chartier 1992, 69–94.
5 Mignolo 2008, 1737–1738.

6 Dussel 1992, 12: "*l'expérience de la domination du centre sur une périphérie*" (my translation).
7 Green, John, *A New General Collection of Voyages and Travels*, London: Thomas Astley, 1745–1747.
8 See Mas 1985, 398–99.
9 On Prévost's life, see Sgard 1989 and Sgard 2013.
10 Crone and Skelton 1967, 98–109; Crone 1949, 85–91; Crone 1951, 69–70.
11 It should be noted that the text in the English version is laid out in two columns as opposed to one in the French version.
12 On this distinction, see Eche 2017, 125–126.
13 Green, *A New General Collection of Voyages and Travels*, I, vii, xi. See Crone and Skelton 1967.
14 Green, *A New General Collection of Voyages and Travels*, I, vii.
15 See Eche 2017, 31–44.
16 See the review in *Journal des Sçavans* (Paris: G.F. Quillau, mai 1746), 771–801.
17 Laugier, *Manière de bien juger des ouvrages de peinture*, 86–87.
18 Proyart, *Histoire de Loango, Kakongo, et autres royaumes d'Afrique*, 318–342.
19 Thompson 2011, 67.
20 Hartog 2001, 356–364.
21 Greenblatt 1991; Daston and Park 1998, 21–66; Rubiès 1999, 76–77; Gomez-Géraud 2000, 50.
22 Mollat du Jourdain 2005, 10.
23 Mollat du Jourdain 2005, 115. See also, Pagden 1993, 10.
24 Urbain 1998, 31.
25 Urbain 1998.
26 Pagden 1993, 5.
27 Ouellet 2015, 9.
28 Wolfzettel 1996.
29 Gomez-Géraud 2000, 50.
30 Sprat, *History of the Royal Society*, 214–215.
31 Kenny 1998, 15.
32 Daston and Park 1998, 66.
33 McGregor 2007, 44–52. See also, Daston and Park 1998, 215–254.
34 Kenny 2011, 357–369.
35 For instance, Thevenot, *Relation de divers voyages curieux*; Biron, *Curiositez de la nature et de l'art*.
36 *Dictionnaire de l'Académie française*, vol. I, 299: "*passion, désir, empressement de voir, d'apprendre, de posséder des choses rares, singulières, nouvelles*" (my translation).
37 Abbé Prévost, *Histoire générale des voyages*, title page: "*Histoire générale des voyages… contenant ce qu'il y a de plus remarquable, de plus utile utile et de mieux avéré*".
38 *Dictionnaire de l'Académie française*, vol. I: "*CURIEUX se dit aussi Des choses, & signifie; Rare, nouveau, extraordinaire, excellent dans son genre. Cette remarque est curieuse. Un bijou curieux. Travail curieux. On dit, qu' Un livre est curieux, qu' un cabinet est curieux, pour dire, qu' Un livre, qu' un cabinet est rempli de choses rares & curieuses*".
39 Abbé Prévost, *Histoire générale des voyages*, XIII, 473–474. Roman figures indicate volumes, Arabic numbers indicate pages.
40 La Condamine, *Journal du voyage fait par ordre du roi a l'Equateur*, 14–15.
41 Abbé Prévost, *Histoire générale des voyages*, XIII, 474.
42 Batten 1978, 58.
43 Green, *A New General Collection of Voyages and Travels*, II, 256–257. Roman figures indicate volumes, Arabic numbers indicate pages.
44 Green, *A New General Collection of Voyages and Travels*, II, 188–189.

45 Abbé Prévost, *Histoire générale des voyages*, II, 46.
46 Green, *A New General Collection of Voyages and Travels*, I, 548.
47 Green, *A New General Collection of Voyages and Travels*, I, p. 553.
48 Béthencourt, *Le Canarien: livre de la conquête et conversions des Canaries (1402–1422) publié d'après le manuscrit original avec introduction et notes* (Société de l'histoire de Normandie, 1874).
49 Castillo 1992, 57–69. Nothing in Scory's text suggests that he used Thomas Nichols's *A Pleasant Description of the Fortunate Ilandes, Called the Ilands of Canaria Vvith their Straunge Fruits and Commodities. Verie Delectable to Read, to the Praise of God. Composed by the Poore Pilgrime*. (London, 1583).
50 Green, *A New General Collection of Voyages and Travels*, I, 549. The last sentence is exactly word for word the one found in Purchas's *Pilgrimage* (London 1613), Book 7, chap. 12, 785.
51 Green, *A New General Collection of Voyages and Travels*, I, 549.
52 Abbé Prévost, *Histoire générale des voyages*, II, 252.
53 Green, *A New General Collection of Voyages and Travels*, I, 555; Abbé Prévost, *Histoire générale des voyages*, II, 259.
54 Green, *A New General Collection of Voyages and Travels*, I, 555.
55 Abbé Prévost, *Histoire générale des voyages*, II, 261.
56 Green, *A New General Collection of Voyages and Travels*, I, 555.
57 Abbé Prévost, *Histoire générale des voyages*, II, 261.
58 The same can be said about La Condamine's example mentioned above: what struck him was the similarity between a South-American cultivated landscape and a French one. The surprise comes from the realization of this similitude in spite of a preconceived set of ideas.
59 Daston and Park 1998, 13.

Bibliography

Primary sources

Béthencourt, Jean De. *Le Canarien: livre de la conquête et conversions des Canaries (1402–1422)*. Société de l'histoire de Normandie, 1874.

Biron, Claude. *Curiositez de la nature et de l'art, aportées dans deux voyages des Indes; l'un aux Indes d'Occident en 1698 et 1699, et l'autre aux Indes d'Orient en 1701 et 1702. Avec une relation abregée de ces deux voyages*. Jean Moreau, 1703.

Dictionnaire de l'Académie française Vol.1. Coignard, 1694.

Green, John. *A New General Collection of Voyages and Travels*. Thomas Astley, 1745–1747.

Journal des Sçavans, Paris: G.F. Quillau, May 1746.

La Condamine, Charles Marie de. *Journal du voyage fait par ordre du roi a l'Equateur*. Imprimerie royale, 1751.

Laugier, Marc-Antoine. *Manière de bien juger des ouvrages de peinture*. Claude-Antoine Jombert, 1771.

Nichols, Thomas. *A Pleasant Description of the Fortunate Ilandes, Called the Ilands of Canaria Vvith their Straunge Fruits and Commodities. Verie Delectable to Read, to the Praise of God. Composed by the Poore Pilgrime*. London, 1583.

Prévost, Antoine-François. *Histoire générale des voyages*. Didot, 1746–1759.

Proyart, Liévin-Bonaventure. *Histoire de Loango, Kakongo, et autres royaumes d'Afrique : rédigée d'après les mémoires des préfets apostoliques de la Mission françoise*. C.P. Berton, 1776.

Purchas, Samuel. *Purchas His Pilgrimage: or Relations of the World and the Religions observed in all Ages and Places discovered, from the Creation unto this Present.* William Stansby for Henrie Fetherstone 1613.

Sprat, Thomas. *History of the Royal Society*. T.R. for J. Martyn at the Bell, 1667.

Thevenot, Melchisédech. *Relation de divers voyages curieux qui n'ont point esté publiées ou qui ont esté traduites d'Hacluyt, de Purchas et d'autres voyageurs anglois, hollandois, portugais, allemands, espagnols et de quelques persans, arabes et autres auteurs orientaux*. Jacques Langlois, 1663–1696.

Literature

Batten, Charles. *Pleasurable Instruction. Form and Convention in Eighteenth-Century Travel Literature*. University of California Press, 1978.

Castillo, Francisco Javier. "The English Renaissance and the Canary Islands: Thomas Nichols and Edmund Scory," in *Proceedings of the Spanish Society for English Renaissance Studies*, edited by Ralph Davies, Universidad of Oviedo, 1992: 57–69.

Chartier, Roger. *L'Ordre des livres. Lecteurs, auteurs, bibliothèques en Europe entre XIVe et XVIIIe siècle*. Alinéa, 1992.

Crone, Gabriel and Skelton, R. "English collections of Voyages and Travels, 1625-1846." in *Richard Hakluyt and his successors*, edited by Edward Lynam. Klaus Reprint, 1967.

Crone, Gabriel. "John Green. Notes on a Neglected Eighteenth Century Geographer and Cartographer." *Imago Mundi*, no.6 (1949): 85–91.

Crone, Gabriel. "Further Notes on Bradock Mead, Alias John Green, an Eighteenth-Century Cartographer." *Imago Mundi*, no.8 (1951): 69–70.

Daston, Lorraine, and Park, Katharine. *Wonders and the Order of Nature*. New-York, 1998.

Dussel, Enrique. *1492 L'occultation de l'autre*. Translated by Christian Rudel. Les Editions ouvrières, 1992.

Eche, Antoine. *Représentations de l'alérité dans l'Histoire générale des voyages. Traduction, récritures et illustrations*. Le Manuscrit, 2017.

Gomez-Géraud, Marie-Christine. *Ecrire le voyage au XVIe siècle en France*. PUF, 2000.

Greenblatt, Stephen. *Marvelous Possessions. The Wonder of the New World.* The University of Chicago Press, 1991.

Hartog, François. *Le Miroir d'Hérodote. Essai sur la représentation de l'autre*, Gallimard, 2001.

Kenny, Neil. *Curiosity in early Modern Europe Word Histories*. Harrassowitz Verlag, 1998.

Kenny, Neil. "La Collection comme mode discursif dans les relations de voyage françaises aux XVIe et XVIIe siècles." *French Studies: A Quarterly Review* 65, no. 3 (2011): 357–369. https://muse.jhu.edu/article/444930.

Mas, Jean-Paul. "Introduction." In *Œuvres de Prévost*, Vol.7, edited by Jean Sgard. Presses universitaires de Grenoble, 1985.

McGregor, Arthur. *Curiosity and the Enlightenment: Collectors and Collections from the Sixteenth to the Nineteenth Century*. Yale University Press, 2007.

Mignolo, Walter. "Racism as We Sense It Today." *PMLA*, 123, no.5 (2008): 1737–1738.

Mollat du Jourdain, Michel. *Les explorateurs du XIIIe au XVIe siècle. Premiers regards sur des mondes nouveaux*. Éditions du C.T.H.S, 2005.

Ouellet, Réal. *La Relation de voyage en Amérique (XVIe-XVIIIe siècles). Au carrefour des genres*. Hermann, 2015.

Pagden, Anthony. *European Encounters with the New World.* Yale University Press, 1993.

Rubiès, Joan-Pau. "Futility in the New World: Narratives of Travel in Sixteenth-Century America.", in *Voyages and Visions. Towards a Cultural History of Travel*, edited by Jas Elsner and Joan-Pau Rubiès. Reaktion Books, 1999.

Sgard, Jean. *Prévost romancier*. Corti, 1989.

Sgard, Jean. *Vie de Prévost (1697–1763)*. Hermann, 2013.

Stagl, Justin. *A History of* Curiosity. *The Theory of Travel-1550–1800*. Routledge, 1995.

Thompson, Carl. *Travel Writing*. Routledge, 2011.

Urbain, Jean-Didier. *Secrets de voyages : Menteurs, imposteurs, et autres voyageurs invisibles*. Payot, 1998.

Wolfzettel, Friedrich. *Le discours du voyageur. Pour une histoire littéraire du récit de voyage en France, du moyen âge au XVIIIe siècle*. PUF, 1996.

10

WONDERS OF THE NIGHT

Nocturnal darkness as a sensory experience

Susanna Lahtinen

This chapter explores nocturnal darkness in 18th- and 19th-century travel writing, focusing on the emotions that night-time aroused in travellers. This approach to the history of travel sheds new light not only on past emotional and sensory experiences but also on the cultural history of darkness. I argue that darkness was a provider of and a background to many experiences of wonder as it intensified the feeling of amazement and admiration that travellers experienced in front of something beautiful, exceptional and unfamiliar. The chapter focuses on the experiences of night-time travel of four British women, Mary Wortley Montagu (1689–1762), Hester Lynch Thrale Piozzi (1741–1821), Charlotte Anne Eaton (1788–1859) and her sister Jane Waldie Watts (1793–1826). In line with Barbara H. Rosenwein's approach to the history of emotions, Montagu, Piozzi, Eaton and Watts are studied as a specific emotional community marked by their gender and upper-class Britishness.

This chapter is based on my dissertation, in which I inquire into darkness as an entity with agency, into the emotions it aroused among past travellers and writers, and into the role of the senses in the process of experiencing the night. The chapter addresses questions such as: What kind of emotions travellers felt towards the nocturnal milieu? What sort of sensory experience was the night? Did nocturnal sights and sceneries arouse wonder? In what follows, I will briefly outline some aspects of the history of nocturnal darkness and the history of travel and travel literature. I will also discuss different approaches to the history of emotions, addressing the specific research questions mentioned above.

DOI: 10.4324/9781003506928-10

Emotions and experiences

This chapter takes its cue from the phenomenological approach to experience. In line with phenomenology, the chapter takes for granted that experience contains a subject and an object and, more specifically, that there is a subject which perceives and an object towards which the conscious activity of perceiving is directed. Experience originates from this. Following the aim of phenomenological philosophy to describe a phenomenon as it appears to the subject, the chapter explores the meaning and significance of lived experiences. Experience can be seen as an umbrella term denoting the myriad ways in which an individual knows the world. As the cultural historian Riitta Laitinen has argued, our world of experience and our experiences are always culturally determined. Our cultural stock of meanings and concepts is present in every situation we encounter.[1]

An increasing number of historians have demonstrated that emotions (passions, sentiments, feelings) have a history. The study of past emotions is often linked to the history of the senses. Cultural historians Bruce Johnson and Hannu Salmi have pointed out that the research field relating to the study of the history of the senses is extremely broad because much of the information about the past is mediated by human perception. They also connect the history of the senses to that of travel and discuss the meaning of sensory experiences, for example, when travellers encounter new smells and tastes in a foreign environment. Furthermore, they suggest that in such encounters the senses participate in the shaping of mental images and argue that the history of the senses is entangled with the history of the emotions.[2] Johnson and Salmi acknowledge that, as a research topic, the sensory environment has its challenges. Inquiry into the sensory impression itself must take into account the possible limitations posed on the senses by the environment. While sometimes historical source material can help us conclude what kind of sensory stimuli were available in a certain environment, this sort of historical reconstruction is not always possible.[3]

The historian Rob Boddice has pointed out certain challenges that a historian of emotions will inevitably face. To begin with, he argues that the word "emotion" is at once intriguing and infuriating because it eludes precise definition, and yet using the word leads to its precise definition. Furthermore, while "emotion" does cover, for example, phenomena relating to expression, it fails to explain sensory experience and the intermingling of the content of consciousness, reason, and mental movements, limiting the field of research to emotions that we are already familiar with.[4] In this chapter I explore words that indicate different sensory experiences towards the nocturnal darkness and delineate the emotions that resulted from such incidents.

The sensory hierarchy changes at night. That nocturnal experience is predominantly aural is reflected in the old Scottish saying: "*The day has eyes, the night has ears*". At night, one must rely on senses other than sight: on hearing, touch, and smell. Some centuries ago, the clatter of horses' hooves on the cobblestones, the ringing of church bells and the barking of dogs helped to navigate in the darkness of the night.[5]

Sensory experiences can be hard to define, since definitions of senses are culturally bound. Some cultures are believed to have more senses than others. And even viewed from the present-day point of view, some senses are more difficult to categorise than others. For instance, the sense of touch is quite challenging to explain, since in addition to tactile feeling, it can also include aspects such as gut feeling and sense of balance.

The hierarchy between the senses can also change when a traveller becomes exposed to a new environment and unfamiliar sensory stimuli. The senses have a communicative role; they provide stimuli from the surrounding world, and these stimuli act as a kind of message. It is significant that the senses adapt themselves to stimuli: the sound you hear every day will no longer be registered, nor will the smell you are already used to. While the communicative element still exists, the adaptation of the senses means that the communication becomes ignored and is no longer described in the sources. Salmi has posed the question whether this means that the communicative meaning has been lost, or it has simply become unnecessary to explain or describe it. When considering the question, Salmi uses the methodical approach of looking for breaking points in the adaptation of the senses, in other words, situations in which some new factor breaks the familiar system in the world of the senses.[6] When, for example, travel writers leave familiar sensory stimuli unarticulated, this may be because the ongoing adaptation of the senses has temporarily confused or impaired the senses. Unusual stimuli, however, are more likely to be noticed and recorded in writing. Since the traveller is usually in a new environment and since the night too represents (to a certain extent) unfamiliarity, it follows that a nocturnal traveller is in a particularly foreign and unfamiliar situation. This suggests that any travel writing that deals with night-time is sensitive to sensory stimuli and the ensuing, emotionally charged language.

The historian Barbara H. Rosenwein defines her concept of emotional communities as "social groups that adhere to the same valuation of emotions and how they should be expressed" and as "groups of people animated by common or similar interests, values, end emotional styles and valuations".[7] She specifies that emotional communities are not always "emotional", they merely share important norms concerning the emotions and the modes of expressing them. Emotional communities can be defined broadly (for example, English upper-class society in the 19th century).[8]

According to Rosenwein, emotions are inchoate until they are given names. Emotional vocabularies are exceptionally important for the ways in which people understand, express, and indeed "feel" their emotions. Words cannot quite compass, for example, what music evokes in us. Our vocabulary may appear inadequate to describe what we feel.[9] Historians of emotions must also bear in mind that sometimes people feign their feelings. According to Rosenwein, feigning indicates the emotional norms of the community. She also guides the historian of the emotions to pay attention to context and word order when reconstructing emotional experiences.[10]

Darkness and night

The night is both familiar and unfamiliar: we know that night is dark, and we know that darkness stands for the unknown. Darkness falls surreptitiously and obscures the landscape and dims our vision. Darkness hides and conceals, creates shadows. Yet, as one historian of night, Roger Ekirch, has put it, "the darkness of the night is palpable".[11] Night stands for the absence of light and the opposite of day. As such, it determines the length of daylight.[12] Night is familiar, but it is also unknown and obscure. Arguably, people have associated different beliefs and symbolism with night because of its unknown, mysterious qualities. The night has always represented chaos, the world of dreams, the time when ghosts and demons become active.[13] Day and night or light and darkness have been firmly set on opposite sides as two elements that incessantly fight each other.[14] Night has been seen as a border, as a final frontier to be conquered.[15] According to Ekirch, night-time has remained a *terra ingognita* of peripheral concern, the forgotten half of human experience, even though families spend long hours in the obscurity that it creates.[16] In the early modern period, night-time was more than a mere backdrop to daily existence as it embodied a distinct culture, with many characteristic customs and rituals. In Britain and America, the darkness was highlighted by referring to it as the "night season".[17]

According to Ekirch, "night was man's first necessary evil, our oldest and most haunting terror". What if the sun did not rise again? The prospect of total, eternal darkness was frightening.[18] Is this to be understood as fear of the dark or fear of things that might be hidden in the dark? While fear of the dark is common among both children and adults, it does not always concern darkness itself. Nyctophobia is characterised by a severe fear of the night. It is triggered by the brain's disfigured perception of what would, or could, happen in a nocturnal environment. What frightens people who suffer from nyctophobia is the possible or imagined dangers concealed by darkness. This is why imagination and beliefs have such a powerful impact on what we experience at night. In his history of night-time, Ekirch argues that night dramatically transformed the communal landscape as it invests innocuous landmarks with sinister portent.[19] Therefore, in darkness, our imagination starts to run wild, and we begin to believe in the unbelievable.

While night-time tends to guide our imagination towards what is dangerous and frightful, people have not always been afraid of darkness *per se*. Like daylight, darkness was considered a natural part of life. It provided privacy, a space where one could be alone and protected from prying eyes.[20] For many, darkness represented a sanctuary from ordinary existence.

It is difficult to understand and navigate darkness without light. While light enables us to move in the dark, it only illuminates a small radius and, in so doing, effectively obscures the surroundings. A small light source often only emphasises the darkness around it. Both artificial and celestial light, with the mystical splendour they create when illuminating the night, are essential factors which shape the ways we experience the aesthetics of darkness. Historian Craig Koslofsky has stated that in the early modern period, nocturnalisation constituted a revolution as urban darkness was overcome with artificial light, and sunset no longer signalled the ending of social activities. The increase in street lighting across cities and, to some extent, in the countryside, made curfew redundant.[21] At the same time, designers of artificial lighting had the power to direct the gaze of after-dark flaneurs and merrymakers. Along the same lines, in a discussion about the development of nocturnal sociability and ceremonial light in 17th- and 18th-century Europe, historian Panu Savolainen describes how, with the help of light, the night gradually became an extension of the day, and fun and new forms of sociability extended into the dark hours of the day and merged into everyday life in cities.[22] Artificial lighting played an equally important role in the development of sightseeing.

Travelling at night, however, depended on the light of the moon and stars. While stellar illumination was weaker than the lunar one, it was nevertheless more important and reliable because stars facilitated navigation. Clouds too played a significant role, as they regulated the flow of celestial light.[23] In cities, a traveller could hire an assistant who illuminated the way with a torch or a lantern. These linkboys were also known as the moon cursers, because of their antipathy towards the moon, whom they considered a rival light bearer and a threat to their livelihood.[24] For example, in Finland, some towns or cities had regulations that prohibited anyone from walking in the dark without a lantern.[25]

Travel and travel literature

When we travel, we always move somewhere, towards some place. The journey involves moving away from everyday routines. We also move around travel destinations because we want to see as much as possible. However, not all geographical movement is tourism, but an activity which is best described as mobility. The term "travel" is often used as a general term for all kinds of mobility, although the term "tourism" often appears alongside it. "Tourism" and "tourist" came into use in the English language at the end of the 18th century initially functioning as synonyms for travel and traveller, but very soon they acquired negative connotations. Already in the middle of the 19th century, people wanted to identify themselves as travellers and not as tourists. This distinction is also clearly emphasised in travel literature.[26] In the history of British tourism, the end of the 18th century was a significant time, as mobility was made easier and terminology related to tourism in general became established. Road networks were improved, and maps became better, making travel easier and safer.[27]

The development of train travel affected the experience of travel, and according to Wolfgang Schivelbusch, the peak of detailed travel was in the 18th century, but with the emergence of railroads, the era of attention to detail ended. Before the fast train travel, travellers had more time to admire sceneries and individual details.[28] In the past, travellers felt like they were part of the landscape, which in turn magnified the emotions they experienced during their journey. Sitting on a train and watching the scenery pass by signalled a change to that kind of detailed travel experience. Before the trains, travelling was a slow and complicated affair: in difficult terrain, when carriages could not move forward, mules carried the carriages and porters carried the travellers in separate litter chairs.[29] For example, the travellers I studied in this article travelled with wagons and carts.

My source material consists of travel literature and letters, which were sent from abroad and often later published as a book. While darkness at a familiar home environment felt safer and was unlikely to be mentioned in letters and diaries, this was not the case in the context of travel when the traveller was in a foreign place and paid attention to the surroundings in a more alert way. This is why travellers' written documents form a particularly apt source base for study on nocturnal experiences.

When researching experiences, the object is a person's lived experience, which took place in the past, and the researcher can only access a description of it.[30] By writing, people make meaning of the things they have experienced. Nothing would have been written in a journal or a letter unless it had some meaning for the writer. Travel diaries are written both at travel destinations and much later, at home when reminiscing about a journey. Because of this, the experiencer of the events, even in autobiographical texts, is no longer the same as she was at the time of the experience. A long time may have passed between the experience and the moment of writing it down, and, what is more, the memory of the experience may have changed. Perhaps the travel writer has grown old and forgetful. Perhaps new information on the subject, learnt from books or otherwise, has started to colour the memories.[31]

A further challenge to the study of past experiences comes from the fact that many travelogues have been printed in several editions, which have caused alterations to the texts. Sometimes in the prefaces of travelogues, the editor of the book mentions that he has shortened or removed passages in the text. This is unfortunate because sections deleted due to repetition, for example, could have revealed something about the significance of the phenomena discussed in them.

From the middle of the 18th century, travel literature was a lucrative, prestigious, and popular genre of literature in Europe.[32] In the early 18th century, by contrast, descriptions of the traveller's own, individual experiences were almost non-existent in travel literature because matter-of-fact prose was preferred.[33] This meant that individual observations and subjective experiences were usually absent in travelogues. Still, it has been pointed out that travel literature has always teetered on the border between private and public. Even if the travel diary or narrative was meant to be an accurate description of the world, it has always included, either consciously or unconsciously, a certain kind of self-image, a personal and culture-bound subjective perspective. Writing about travel in all its manifestations always contains some kind of autobiography.[34] That travel writers always reveal, intentionally or not, their own impressions about things they perceive and experience makes their writing excellent material for research on past experience.

A big change took place in the era of Romanticism, the artistic and intellectual movement that started in the late 18th-century Europe and focused on emphasising emotions and individualism. From that time onwards, it was appropriate to bring out one's own personality and experiences in writing. Furthermore, during the Romantic period, travel destinations became more diverse, familiar routes began to be described in a new way, and conventions were liberalised.[35]

However, learnt practices and standards of each period had an impact on people's emotions and feelings. Much depended on what was considered to be standard and acceptable, what one could say or write without being judged. If some feelings or emotions were not allowed to be demonstrated, this no doubt affected the ways in which people expressed themselves, let alone described their experiences in a letter or a journal. Furthermore, what was considered beautiful and worthy of admiration changed with time. When 18th-century philosopher Edmund Burke wrote his book *A Philosophical Enquiry into the Origin of Our Ideas of the Sublime and Beautiful* in 1757, his ideas about the sublime, for example, gave travellers novel ways to describe impressive sceneries. The sublime landscape included dramatic views, a sense of infinity, and variations of light and shadow.[36] According to Burke, darkness remained always terrible in its own nature.[37] What is more, he states that darkness evokes greater emotions than its opposite, light.[38] Fear is a pivotal element in experiencing strong emotions, and without a strong reaction, no experience can be sublime and evoke emotions such as admiration, respect, and astonishment.[39]

As stated above, travel writing has its challenges as source material. Travel diaries and reports were written with the readership in mind and according to the conventions of the time. The act of admiring sights was culturally determined and resembled a predetermined ritual: certain destinations just had to be seen.[40] Travel literature is characterised by repetition; many travel documents refer to similar experiences, and the same comments are repeated regardless of the person who experienced them.[41] I argue that by studying the emotions that were evoked by nocturnal milieu, we can find experiences and expressions that reveal novel and less repetitive aspects of early modern travel writing.

In general, travellers used similar language and shared certain conventions of the genre of travel writing when recording their experiences. What is more, they often travelled to the same destinations and used the same routes. Furthermore, as we have seen, it was customary for travel writers to reiterate identical experiences in similar fashion and thus create and sustain a convention. While the following discussion on Montagu, Eaton, Waldie, and Piozzi explores their emotionally charged reflections regarding nocturnal darkness, I delve into similarities and differences in their observations in order to conclude what sort of sensory experience was the night? What kind of emotions travellers felt towards nocturnal milieu? Did nocturnal sights and sceneries arouse wonder?

Travellers' nocturnal experiences

An English aristocrat and poet, Lady Mary Wortley Montagu (née Pierrepont; 1689–1762) wrote a number of letters while travelling in regions that are now parts of Germany, Turkey, and Tunis. Her correspondence was published posthumously in several volumes. In this study I use *Letters of Lady Mary Wortley Montague*, written during her travels in Europe, Asia, and Africa: to which are added poems by the same author, published in 1817. Hester Lynch Thrale Piozzi (née Salusbury 1741–1821) wrote a travelogue that describes her honeymoon travels, which was published as *Observations and Reflections made in the Course of a Journey through France, Italy, and Germany* in 1789 (in this study I use volume 1). Jane Waldie Watts (née Waldie 1793–1826) wrote *Sketches Descriptive of Italy*, published in 1820 (again, in this study I use volume 1). Her sister Charlotte Anne Eaton (née Waldie; 1788–1859) was a travel writer and novelist who later refashioned herself as a banker. Eaton's travel writing covers her journeys to Belgium and Rome. In 1820, she published her letters from Rome as the book *Rome in the Nineteenth Century: containing a complete account of the ruins of the ancient city, the remains of the Middle Ages and the monuments of modern times: with remarks on the fine arts, on the state of society, and on the religious ceremonies, manners and customs of the modern Romans*. As the title suggests, Eaton's writing was influenced by Romanticism and its characteristic tendency to glorify the past. In line with Romanticism, her travel writing is marked by intense emotion as an authentic source of aesthetic experience. Sympathy, awe, wonder, and terror number along key elements of travel writing of the period, and they were used to express emotional responses, for example, to the beautiful and the sublime. She shares the same romantic style of writing with her sister.

In contrast to Romanticism, in the period when Montagu travelled, it was more common to describe what one saw without reflecting on one's emotions. In general, in Montagu's period, it was more customary to write reports in which chronological narrative was combined with topographical and ethnographic descriptions. Since travel was to educate travellers and the readers of travel books, a specific genre of instructions for the scientific traveller started to emerge in the 17th century.[42]

Piozzi, as a traveller between Montagu and the sisters Eaton and Waldie, was not yet a product of the age of Romanticism. Indeed, her writing style differs from the latter two as it lacks the Romantic, overflowing praise for everything that has been perceived and experienced. At the same time, her descriptions are perceptive, and she uses beautiful turns of phrase. Although writing before the age of Romanticism, there are also beautiful descriptions of personal experiences in Montagu's travelogue.

Montagu, Eaton, Waldie, and Piozzi travelled in company. Montagu and Piozzi travelled with their husbands, Eaton and Waldie with each other (and, for a while, with their brother). It is likely that having a travel companion changed the nature of their voyage: alone, travelling through a dark landscape or admiring sights in the night would have been a different experience.

Why did people travel by night in the early modern period? Sometimes this was a choice between a night spent in an unpleasant, foul-smelling inn, or in a carriage that continued its journey through the dark landscape. In a letter composed in Germany, Montagu laments the untidiness of the places she encountered and (although she had packed her own bed with her) states that she preferred to travel through the night, wrapped in furs to stay warm.[43] This suggests that, at least sometimes, travelling by night was considered a less uncomfortable option than an overnight stay at an inn.

Excessive heat of the day was another reason why night was sometimes the preferred time to travel. This is what happened to Montagu in Tunis:

> "I set out in his chaise at nine at night, the moon being at full. I saw the prospect of the country almost as well as I could have done by day-light, and the heat of the sun is now so intolerable, 'tis impossible to travel at any other time".[44]

Since the full moon illuminated the night, Montagu and her travel companions spent two hours admiring with great attention the remains of a noble aqueduct, which was situated six miles from Tunis. Montagu writes that it had many complete arches and, in the past, it had carried water to Carthage, over several high mountains, the length of forty miles. She also notes that, according to her husband, the aqueduct they had seen in Rome was inferior to the one before them. After this nocturnal sightseeing, they arrived at Tunis soon after daybreak.[45] While the aqueduct was great to be admired in cool night air and in the light of the moon, her morning visit to the ruins of Carthage proved a less pleasant experience because of the excessive heat.[46] When visiting the location of the ancient Troy, Montagu made the effort of waking up at two at night to see the ruins in cool air:

> "We sailed next night to the shore where 'tis vulgarly reported Troy stood, and I took the pains rising at two in the morning, to view coolly those ruins which are commonly shewed to strangers".[47]

Besides being too hot for travelling, daytime in Tunis was also too bright for writing, Montagu points out. In one of her letters, she apologises for not having written from Tunis:

> "The heat there was so excessive, and the light so bad for the sight, I was half blind by writing one letter to the abbot -, and durst not go to write many others I had designed".[48]

On the other hand, Eaton observed that it was sometimes difficult to write at night. She remarks in one of her travelogue entries: "But I must take up the history of our adventures where I left them off, when the shades of evening stopped my carriage epistle".[49]

Waldie mentions a light they used when travelling in the dark:

> "About five in the evening we arrived at Macon, where we walked about the town, dined, and slept; and next morning re-entered the boat before five, by candle, or rather lantern-light. When we set out, it was perfectly dark; but the moment the dawn did commence, it was almost instantly day".[50]

The lantern is mentioned again in another passage:

> "—were all seated in our berline by half-past four, and fairly off by the light of a paper lanthorn, suspended in front of the ponderous machine, giving no bad idea of the one-eyed Polythemus; or, if my reader prefer a less classical simile, he may bethink himself of an English stage-waggon that travels at night".[51]

Eaton describes the coolness of the evening and freshness of the many fountains as a luxury that can only be appreciated by someone who has experienced the heat of a scorching day in a warm climate:

> "It is a luxury, the full value of which cannot be felt but in such a climate as this; and those only who have known that delicious moment, when the blaze of the summer-day fades at last in the golden clouds of evening, can

understand the voluptuous delights with which, in its hushed hour of stillness and repose, you listen to the music of their dashing murmur, and rest beside their freshness".[52]

Piozzi does not seem to write about the heat of the southern climate as much as, for instance, Montagu or Eaton, and she only refers to it when describing trees that could provide shelter. It is also possible that the heat in Italy, where Piozzi travelled, was not as extreme as in Tunis, for example.

Sometimes we can find feelings and expressions of disappointment in travelogues. This is the case when a traveller, for example, saw something for the first time and the sight did not match the expectations. The horizon produced a sort of anti-wonder for Piozzi when she, for the first time, saw the sun setting and rising without anything obscuring the view:

> "Seeing the sun rise and set, however, upon an unobstructed horizon, was a new idea gained to me, who never till now had the opportunity. It confirmed the truth of that maxim which tells us, that the human mind must have something left to supply for itself on the fight of all sublunary objects. When my eyes have watched the rising or setting sun through a thick crowd of intervening trees, or seen it sink gradually behind a hill which obstructed my closer observation, fancy has always painted the full view finer than at last I found it".[53]

Montagu's travel writing lacks Eaton's passionate admiration for the splendour of the moon and moonlit scenery. This difference reflects a change in style and taste between early 18th century and Romanticism and suggests that, despite the many similarities in their outlook, Montagu and Eaton represented different emotional communities. Waldie naturally fits in the same emotional community as her sister. Piozzi, on the other hand, is closer to Eaton and Waldie, even though her writing lacks their ardent passion towards celestial light. This was due to certain changes in culture rather than to their nationality, gender, or class. For example, the following quotation from Eaton differs drastically from the style of Montagu:

> "we beheld the magnificent spectacle of the sun setting in a flood of glory; while the beautiful star of evening lighted her lamp in western sky, and the full glowing moon rose majestically behind the Appenines, to light us on our way through the vales of Tuscany".[54]

Eaton's description of fountains in front of St. Peter's Basilica is equally characteristic of Romanticism:

> "I know not how to describe to you their beauty; but visit them in the repose of evening, when that moon, which here shines like a brighter planet, walks

> in her glory through the heavens; when the stars awake their mysterious fires, and the soft moon beam falls upon the lines of the Grecian columns, ... - visit them then, and you will feel their enchantment".[55]

Also Waldie mentions the moon: "so we were obliged to descend into common life again, and be thankful to the friendly moon, which arose soon after the glow of evening had faded into darkness, 'to light us on our way' not to Mantua, but to the village of Poggibonsi, where we slept".[56] Piozzi too admired sights in the moonlight, as we can judge from her description: "we were driving last night to look at the Colisseo by moon-light there were a few clouds just to break the expanse of azure and shew the gilding".[57] When she was in Venice and beheld the St. Mark's Square in moonlight, it's wondrous lustre made a strong impact: "The general effect produced by such architecture, such painting, such pillars; illuminated as I saw them last night by the moon at full, rising out of the sea, produced an effect like enchantment".[58]

When observed in a foreign sky, the moon appeared more wondrous than back home. Eaton declared that "without any romance, I do assure you the moon does look larger, and shines with far more warmth and brilliancy, in the sky of Italy, than amidst the fogs and vapours of England; a thing by no means unreasonable or unaccountable".[59] What overshadowed the glorious beauty of the moon, however, was the brief period of twilight that follows immediately after the sunset. In that specific light, Eaton asserts, the scenery gained additional charm, something that it did not have at any other time of the day. She adds that the power of such a view was irresistible. In moonlight, there is coldness and unvaried whiteness, she writes, while the short-lived hour of twilight illuminates nature in a state of stillness, harmony, and repose, with softly glowing hues that no pencil could ever equal. While poets preferred to view ruins in pale moonlight, Eaton insisted that the soft shades of twilight made the sight more delightful and dear.[60] Twilight also aroused melancholy, Eaton asserts. Referring to a night void of moonlight, she writes how they travelled through a desolate landscape with no human traces, and as the last rays of light lit up the view, she experienced deep and hopeless melancholy, the like of which she had not felt before.[61] The emerging darkness chimed with Eaton's romantic mindset, guiding her emotionally coloured writing.

When Waldie writes about moonlight, she distinguishes between the impact that such light had on different types of sights:

> "The night was bright and clear, and the moon nearly at the full. By thus driving about to one church after another, I had an opportunity of comparing the principal buildings of Rome under this serene light, which I might perhaps never have seen otherwise. And it is a sight it would be a thousand pities not to behold. The moon-light of southern climates is ten times more beautiful than that we are accustomed to; and where, in the whole world, is

there a spot so suited to aid its lovely effect as Rome? Madame de Stael recommends persons to visit St. Peter's by moon-light. The fountains certainly look beautiful-and so do the colonnades-but I confess, however heretical the opinion may be, that the façade of the church appeared, in my eyes, to much greater advantage by the common day-light. It is 'the ruins gray' that should be visited by 'the pale moon-light'. The desolate arcades of the Coliseum, and the ruined temples of the Forum, are far more suited to this soft and melancholy light, than the bright uninterrupted yellowness of St. Peter's-rich in modern magnificence and ornament".[62]

Describing the challenges of nocturnal travel, Eaton writes how they once travelled more than two hours in total darkness, which made the carriage driver afraid of robbers and the passengers quake because of the cold. When they arrived at the inn, they were all hungry and weary.[63] Also, Waldie mentions the possibility of robbers lurking in the night. She describes how they proceeded on their journey by star-light through a place that had the reputation of being used by robbers and adds that one of her friends had been attacked two years earlier when travelling in the dark.[64]

Similar fears are mentioned by Waldie after a nightly adventure, during which the muleteers were terrified to death by the fear of robbery and other misadventures and swore never to experience that again.[65]

While Eaton and Waldie often travelled by night, it is clear that this was not always pleasurable or without difficulties: "by the time we reached San Secondo", Waldie writes, "evening was beginning to close in; and, as there was some awkwardness about getting the carriage across in the dark when the river was so swelled, we were obliged to remain there all night".[66] As mentioned earlier, darkness can be palpable, and light only illuminates a small radius. This was what Waldie once experienced when travelling at night with her sister: "flashes of very bright lightning occasionally broke upon us, and, bewildering our sight, only served to obscure the light which the clear stars might otherwise have shed over our difficult path".[67] Describing the same journey, she added: "at this time we would willingly have dispensed with its flashes, beautiful as they were, for they served only to mark the 'palpable obscure;' making the intervals appear much darker".[68]

One night, Montagu admired the flames of Etna:

"Alas! art is extinct here; the wonders of nature alone remains; and it was with vast pleasure I observed those of mount Etna, whose flame appears very bright in the night many leagues off at sea, and fills the head with a thousand conjectures".[69]

Montagu was elated before the natural wonder, which she describes as powerful enough to eradicate everything manmade. The contrast between the darkness of the night and the brightness of the flames of the volcano no doubt intensified her experience.

Nocturnal darkness, the brightness of the flames and superstition are mentioned by Waldie, when she describes her brother leaving at one evening to take a closer look at a flame they saw bursting out from the side of a mountain (she did not want to join him, because it was very late and cold). She writes that she "did not much relish the notion of wandering about in the dark on the top of the Apennines", adding, however, that the flame "flashed over dark mountains in fitful splendour" and that she "did not wonder that superstition had fabled it the residence of some malignant demon".[70]

The contrast between light and darkness also animates Eaton's description of her visit to the Colosseum:

> "The still, pale moonbeam fell on the lines of its projecting columns, range above range, to the lofty attic, in silvery light, leaving the black arches in mysterious darkness".[71]

The spectacle of light against the darkness of the night is also discussed in Piozzi's text:

> "The evening concluded, and the night began with fireworks; the church, or duomo, as a cathedral is always called in Italy, was illuminated on the outside, and very beautiful, and very very magnificent was the appearance. The reflection of the cupola's lights in the river gave us back a faint image of what we had been admiring".[72]

The moonlit scenery is a theme that recurs frequently in Eaton's writings. Her admiration for moonlight even led some to consider her "slightly deranged":

> "The best view of the Pantheon is from the windows of this palace. I saw it by moonlight, when the softened light gleamed beautifully in silver lines upon the tall columns of tho [*sic*] portico, more distinctly marked by the dark shade of the intercolumniations, while the serenity of the sky, and the stillness that reigned over every thing, made its beauty more deeply felt. You will wonder what took me to the Giustiniani Palace at night, but we happened to be dining with Count Funchal, the Portuguese ambassador, who lives there; and, as it was moonlight, I took the opportunity of looking out at the Pantheon; ever since which time, his excellency has entertained a rooted conviction that I am slightly deranged; and never sees me without asking, if, when the moon is at the full, I will not return to his house to see the Pantheon. 'Well, that is what I don't understand,' said an Italian Principessa, when he told her of this extraordinary fancy of mine-, for, certainly, one can see plainer in the day-time than the night".[73]

While in Eaton's writing the moon features as one of her most admired sights, her journal reveals that for her, the most sublime terror of nature was produced by a cataract. She writes that it strikes the soul with awe and wonder:

> "To me a mighty cataract has always seemed the most sublime of all the terrors of nature. There is something in its continuity and its unabating rage, which strikes our soul with awe and wonder. All things else in nature change and perish,—and all that are the most fraught with force and power, are the most evanescent, excepting this. The tempests of the ocean pass away,—the thunder-storm endures but for an hour,—the dread hurricane is soon at rest—the volcano's red streams of liquid fire grow cold, and are extinguished—and the earthquake itself, that shakes the foundations of the earth, and swallows up whole nations in its yawning womb, is but the convulsion of a day. But we behold the ceaseless fall of that torrent, which has held on its raging course from the beginning of time, and will continue till its latest close,—which knows no rest, no stop, no change,—by night and by day, in storm and in sunshine, the same in every moment of the past and the future—yesterday, to-day, and for ever!"[74]

Further expressions of wonder towards nocturnal nature are present in Piozzi's observation that evening "shuts in strangely quick in Tuscany". Furthermore, she describes "some unaccountable sparks of fire that seemed to strike up and down the hedges". She was later told that the origin of these small fires in the night was lucciola or fireflies, which flicker at twilight.[75]

The destination of one's journey has sometimes meant that the night was not dark at all. Scandinavian summer nights have aroused amazement and admiration among myriads of travellers. Mary Wollstonecraft (1759–1797), a writer and advocate of women's rights, travelled with her little baby and a servant in Scandinavia in the last years of the 18th century. She describes a summer night that could hardly be called a night because its light competed with the day. During such a night, she adds, one can write well without a candle.[76]

Night has also represented solitude and silence. At the same time, the stars and the moon, along with the view of the ruins at night, made Eaton appreciate the past and contemplate that the same lights of heaven had shone to the people who had lived before her. When visiting the Colosseum, Eaton writes:

> "No sound met our ear but the measured tread of our own footsteps, and the whispered murmurs of our own voices. The deep solitude and silence, the immensity and the ruin of the great fabric that surrounded us, filled our minds with awe; and as we caught the view of the stars appearing and disappearing through the opening arcades, marked the moonbeams illumining the wide range of these lofty walls, and raised our eyes to the beauty of the calm, clear firmament above our head,—we could not but remember that,

> many ages past, these eternal lights of heaven had shone on the sloping sides of this vast amphitheatre when they were crowded with thousands of human beings, impatient for the barbarous sports of the rising day,—where now only the wild weeds waved as the night-breeze passed over them".[77]

Travel documents brim with material that is relevant to historians of emotions. Indeed, Eaton once concluded that she had to stop writing because she was "far too much exhausted from all the sights and sensations".[78] All in all, travellers saw so many wondrous sights to admire that sometimes it seemed to be almost too much to take in, like Eaton once declared:

> "I am so lost in wonder, admiration, and delight, that I know not how to begin. I despair of finding words to describe the objects around me, or give back the faintest image of the various impressions, and the multiplicity of feelings, that crowd upon each other, and overpower me with their force. My mind is confused and agitated like a tumultuous sea, and thought chases thought as rapidly as its waves roll over each other. O that I could transport you here, and make you a sharer in all I see, and feel, and think, and admire; for I can admire, but I cannot describe it!".[79]

Conclusion

What sort of sensory experience was the night? What kind of emotions travellers felt towards nocturnal milieu? Did nocturnal sights and sceneries arouse wonder? Montagu's, Eaton's, Waldie's, and Piozzi's experiences of the night-time related to and depended on sensory perceptions. Darkness was experienced with a number of senses—it was felt, seen, and heard. Eaton, for example, describes the stillness and the silence of the night in her texts. Darkness was also experienced as a whole, as the sum of all the perceptions created by different senses. Montagu felt a sort of delightful rapture at night-time because of the coolness of the night. In a hot climate, the combination of moonlight and cool air made the exploration of new, unfamiliar sights an unforgettable experience. Even mundane activity such as writing acquired a special, wondrous hue in the moonlight.

Expressions of wonder as an emotion that travellers experienced at night are not restricted to descriptions that explicitly and exclusively refer to the word "wonder". As this chapter has sought to demonstrate, Montagu, Eaton, Piozzi, and Waldie experienced wonder while travelling at night, although, looking at their nocturnal experiences from the point of view of the term "wonder", it is evident that the word itself is missing from their descriptions of phenomena that caused astonishment and amazement. It was more typical for the travellers discussed in this chapter to use the term "wonder" when they sought to express something that made them feel curious or doubtful. For example, Montagu uses the word several times, but usually when referring to local customs or antiquities. Eaton also wonders frequently but uses the word precisely to denote thinking about something, and the same goes with Waldie and Piozzi. All of them use the word "admire" when they have seen something wonderful, beautiful, admirable, or wondrous. While the word "wonder" does not occur that often in their descriptions of nocturnal experiences, wonder can still be read in the emotionally charged language that reveals their admiration and wonder towards what they perceived. Nocturnal darkness no doubt produced sublime experiences with mixed feelings of fear, wonder, and admiration.

As mentioned at the beginning of this chapter, darkness was the background. Without nocturnal darkness, the light of the moon, stars, flames of the volcanoes, fireflies, and the light of the lanterns would not have been experienced, or they would have appeared very different. Night-time made an impression on early modern travellers: it enchanted with its magical, melancholy, and serene qualities. Darkness, twilight, and coolness of the night intensified the emotions that Montagu, Eaton, Waldie, and Piozzi experienced in front of the exotic sights they encountered. In such circumstances, volcanoes and ancient ruins appeared particularly wondrous and exceptional to travellers accustomed to different architecture and natural landscapes. On the one hand, nocturnal darkness was difficult, challenging, formidable, and a nuisance when it disturbed travelling. On the other, it was experienced as intriguing, majestic, and sublime with the stars and the moon reflecting their light wondrously against the dark skies while illuminating the landscape.

Notes

1 Laitinen 2004, 1–2.
2 Johnson and Salmi 2013, 86–87.
3 Johnson and Salmi 2013, 91.
4 Boddice 2022, 253.
5 Ekirch 2005, 132.
6 Salmi 2002, 352. See also, Johnson and Salmi 2013, 86.
7 Rosenwein 2016, 3.
8 Rosenwein 2016, 3.
9 Rosenwein 2016, 4.
10 Rosenwein 2016, 6.
11 Ekirch 2005, xxxi.
12 Cabantous 2009, 13.
13 Schivelbusch 1995, 81.
14 Broberg 2016, 35.
15 Cabantous 2009, 13.
16 Ekirch 2005, xxxv.
17 Ekirch 2005, xxv–xxvi.
18 Ekirch 2005, 3.
19 Ekirch 2005, 140.
20 Garnert 2016, 16.
21 Koslofsky 2012, 276.

22 Savolainen 2017, 276.
23 Ekirch 2005, 129–130.
24 Ekirch 2005, 125.
25 Savolainen 2015, 57.
26 Kinsley 2016, 238.
27 Kinsley 2016, 237.
28 Schivelbusch 1996, 50.
29 Black 2011, 34.
30 Perttula 2015, 143.
31 Pöysä 2015, 82.
32 Korte 2016, 173.
33 Bohls and Duncan 2008, xx.
34 Cooke 2016, 15.
35 Bohls and Duncan 2008, xxxiii.
36 Söderholm 2006, 149–150.
37 Burke. *A Philosophical Enquiry*, 115.
38 Burke, *A Philosophical Enquiry*, 65.
39 Burke, *A Philosophical Enquiry*, 45.
40 Koivunen 2006, 121–122.
41 Syrjämaa 2008.
42 Bohls and Duncan 2008, xxiii.
43 Montagu, *Letters*, 42.
44 Montagu, *Letters*, 194. Montagu refers to the English consul whom Montagu and her husband met while staying in Tunis.
45 Montagu, *Letters*, 195.
46 Montagu, *Letters*, 196.
47 Montagu, *Letters*, 189.
48 Montagu, *Letters*, 198.
49 Eaton, *Rome in the Nineteenth Century*, vol. I, 92.
50 Waldie, *Sketches descriptive of Italy*, 26.
51 Waldie, *Sketches descriptive of Italy*, 86–87.
52 Eaton, *Rome in the nineteenth century*, vol. II, 28.
53 Piozzi, *Observations and Reflections*, 2. p.
54 Eaton, *Rome in the Nineteenth Century*, vol. I, 37.
55 Eaton, *Rome in the Nineteenth Century*, vol. II, 29.
56 Waldie, *Sketches descriptive of Italy*, 290.
57 Piozzi, *Observations and Reflections*, 414.
58 Piozzi, *Observations and Reflections*, 151.
59 Eaton, *Rome in the Nineteenth Century*, vol. I, 37.
60 Eaton, *Rome in the Nineteenth Century*, vol. I, 90.
61 Eaton, *Rome in the Nineteenth Century*, vol. I, 68–69.
62 Waldie, *Sketches descriptive of Italy*, 348–349.
63 Eaton, *Rome in the Nineteenth Century*, vol. I, 92.
64 Waldie, *Sketches descriptive of Italy*, 199.
65 Waldie, *Sketches descriptive of Italy*, 166.
66 Waldie, *Sketches descriptive of Italy*, 227.
67 Waldie, *Sketches descriptive of Italy*, 162.
68 Waldie, *Sketches descriptive of Italy*, 163.
69 Montagu, *Letters*, 193.
70 Waldie, *Sketches descriptive of Italy*, 274.
71 Eaton, *Rome in the Nineteenth Century*, vol. II, 390.
72 Piozzi, *Observations and Reflections*, 288–289.
73 Eaton, *Rome in the Nineteenth Century*, vol. II, 123–124.
74 Eaton, *Rome in the Nineteenth Century*, vol. II, 330.

75 Piozzi, *Observations and Reflections*, 307–308.
76 Wollstonecraft, *Letters written during a short residence in Sweden, Norway, and Denmark*, 20.
77 Eaton, *Rome in the Nineteenth Century*, vol. II, 390–391.
78 Eaton, *Rome in the Nineteenth Century*, vol. I, 120.
79 Eaton, *Rome in the Nineteenth Century*, vol. I, 109.

Bibliography

Primary sources

Eaton, Charlotte Anne. *Rome in the Nineteenth Century: containing a complete account of the ruins of the ancient city, the remains of the Middle Ages and the monuments of modern times: with remarks on the fine arts, on the state of society, and on the religious ceremonies, manners and customs of the modern Romans*. Vol. I, second edition. Edinburgh: Archibald Constable, 1822.

Eaton, Charlotte Anne. *Rome, in the nineteenth century; containing a complete account of the ruins of the ancient city, the remains of the Middle Ages, and the monuments of modern times. with remarks on the fine arts, on the state of society, and on the religious ceremonies, manners and customs of the modern Romans*. Vol. II, fifth edition. London: Henry G. Bohn, 1852.

Piozzi, Hester. *Observations and Reflections made in the Course of a Journey through France, Italy, and Germany*. London: A. Strahan; and T. Cadell in the Strand, 1789.

Waldie, Jane. *Sketches descriptive of Italy in the years 1816 and 1817, with a brief account of travels in various parts of France and Switzerland in the same years*. Volume 1. London: John Murray, 1820.

Wollstonecraft, Mary. *Letters written during a short residence in Sweden, Norway, and Denmark*. London: Cassell, 1889.

Wortley Montagu, Mary. *Letters of Lady Mary Wortley Montague, written during her travels in Europe, Asia, and Africa: to which are added poems by the same author*. Paris: P. Didot and F. Didot, 1817.

Literature

Black, Jeremy. *The Grand Tour in the Eighteenth Century*. Gloucestershire: The History Press, 2011.

Boddice, Rob. *Tunteiden historia*. Translated by Jyrki Vainonen. Tampere: Vastapaino, 2022.

Bohls, Elizabeth A. and Ian Duncan. *Travel Writing 1700–1830. An Anthology*. Oxford: Oxford University Press, 2008.

Broberg, Gunnar. *Nattens historia*. Stockholm: Natur & Kultur, 2016.

Burke, Edmund. *A Philosophical Enquiry into the Sublime and Beautiful*. Oxford: Oxford University Press, 2015.

Cabantous, Alain. *Histoire de la Nuit XVIIe-XVIIIe siècle*. Paris: Fayard, 2009.

Cooke, Simon. "Inner Journeys: Travel Writing as Life Writing". In *The Routledge companion to travel writing*, 173–184, edited by Carl Thompson. London: Routledge, 2016.

Ekirch, A. Roger. *At Day's Close. Night in Times Past*. New York: W.W. Norton & Company Ltd., 2005.
Garnert, Jan. *Ut ur mörkret. Ljusets och belysningens kulturhistoria*. Lund: Historiska Media, 2016.
Johnson, Bruce and Hannu Salmi. "Aistien historia: Kohteet ja menetelmät". In *Tulkinnan polkuja. Kulttuurihistorian tutkimusmenetelmiä*, 82–106, edited by Asko Nivala and Rami Mähkä. Turku: Cultural History - Kulttuurihistoria 10, 2013.
Kinsley, Zoe. *Women Writing the Home Tour, 1682-1812*. Hampshire: Ashgate Publishing Limited, 2008.
Koivunen, Leila. "Nähtävyyksiin tutustumassa". In *Turistin tilat. Tilallisuus modernin matkustajan kokemuksissa, 121–134*, edited by Leila Koivunen, Taina Syrjämaa and Ilse Söderholm. Jyväskylä: Turun Historiallinen Yhdistys, 2006.
Korte, Barbara. "Western Travel Writing 1750–1950". In *The Routledge companion to travel writing*, 173–184, edited by Carl Thompson. London: Routledge, 2016.
Koslofsky, Craig. *Evening's Empire. A History of the Night in Early Modern Europe*. New York: Cambridge University Press, 2012.
Laitinen, Riitta. "Johdanto tilan kokemisen kulttuurihistoriaan". In *Tilan kokemisen kulttuurihistoriaa*. Turku: Cultural History – Kulttuurihistoria 4, 2004.
Perttula, Juha. "Kokemus ja kokemuksen tutkimus: fenomenologisen erityistieteen tieteenteoria". In *Kokemuksen tutkimus. Merkitys – Tulkinta – Ymmärtäminen*, edited by Juha Perttulan and Timo Latomaa. Rovaniemi: Lapin yliopistokustannus, 2015.
Pöysä, Jyrki. *Lähiluvun tieto. Näkökulmia kirjoitetun muistelmakerronnan tutkimukseen*. Joensuu: Suomen Kansantietouden Tutkijain Seura, Kultaneito XVII, 2015.
Rosenwein, Barbara H. *Generations of Feeling. A History of Emotions 600–1700*. Cambridge: Cambridge University Press, 2016.
Salmi, Hannu 2002. "Onko tuoksuilla ja äänillä menneisyys? Aistiympäristön historia tutkimuskohteena". In *Kulttuurihistoria. Johdatus tutkimukseen*, edited by Kari Immonen and Maarit Leskelä-Kärki. SKS, 2002.
Savolainen, Panu. "*Yön ainoat valopilkut. Turun katuvalaistuksen alkuvaiheet ja kaupunkiyhteisö*". Turku: Turun seudun sukututkijoiden vuosikirja 57–75, 2015.
Savolainen, Panu. *Teksteistä rakennettu kaupunki. Julkinen ja yksityinen tila turkulaisessa kielenkäytössä ja arkielämässä 1740-1810*. Turku: Bibliotheca Sigillumiana 2, Sigillum, 2017.
Schivelbusch, Wolfgang. *Disenchanted Night. The Industrialization of Light in the Nineteenth Century*. Translated by Angela Davies. London: University of California Press, 1995.
Schivelbusch, Wolfgang. *Junamatkan historia*. Translated by Margit Heinämäki. Jyväskylä: Vastapaino, 1996.
Syrjämaa, Taina. "Aistittu Rooma. Matkailijan koettu ja kerrottu kaupunki". *Ennen ja nyt. Historian tietosanomat*. Vol. 8, Nro 3–4 (2008).
Söderholm, Ilse 2006. "Maiseman äärellä". In *Turistin tilat. Tilallisuus modernin matkustajan kokemuksissa, 121–134*, edited by Leila Koivunen, Taina Syrjämaa and Ilse Söderholm. Jyväskylä: Turun Historiallinen Yhdistys, 2006.

11

COLUMBUS, TRAVEL AND WONDER

A Decolonial Reading from Latin American *Marvelous Real*

María Isabel Gaviria

Introduction: Wonder aesthetics in the context of modernity/coloniality

The marvelous relates to extraordinary experiences that are marked by emotional responses. It is associated with a surprising encounter, with the supernatural, with religious objects and natural phenomena. Its multifaceted nature allows it to manifest itself in various forms that oscillate between a "cognitive and emotional" state and an "observed object or event".[1] In medieval and early modern literature "wonder" was referred to with the Latin word "*admiratio*" at the root of which is the verb "*miror*", to marvel at, to wonder, which has an evident emotional connotation.[2] "*Admiratio*" thus connotes an emotional response and implies a form of contemplation and an active participation in seeing. The words "mirabilia" or "miracula" were used to refer to objects that caused wonder. It seems that the verb "*miror*" and the adjective "*mirus*" have their roots in the Indo-European word for "smile".[3] Revealing the emotional effect and significance of admiration, the etymology of wonder demonstrates how marvel operates. Understanding the full, etymological meaning of "marvel" is important for the purposes of this chapter: in Columbus, it functions as a colonial rhetorical procedure, and in the *Marvelous Real*, it becomes an aesthetic element that defines the identity of Latin American literature.

Wonder is a category that encompasses various phenomena grouped by a feeling of novelty and admiration that appears unexpectedly. While associated with what is surprising, in the Middle Ages and early modern period, wonder was common in everyday life and embedded in the social imaginary.[4] Living with monsters, oddities and precious objects, at least in the imagination,

DOI: 10.4324/9781003506928-11

suggests a culture that connected the material with the textual and evoked a desire to produce wonders as commodities. Stories about wonders were spread through travel literature brimming with depictions of beings from other worlds and descriptions of unknown, magical objects.[5] "These texts created or described objects which were appreciated and thus reflected the taste of their readers".[6] In this way, the books and commodities in circulation both responded to and increased the desire to produce the marvelous, both as condition and materiality.

This familiarity with wonder, associated with a world imagined by travellers describing unknown worlds, helps us understand Columbus' mindset when he arrived in the "Indies". Faced with the novelty and strangeness of this new world and overwhelmed by the challenge of describing it, Columbus appeals to images and commonplaces provided by the imaginary of the medieval and early modern wonder. In the letter to Santángel, the admiral expresses his first impressions of the nature and inhabitants of the new world in a manner that will be decisive for the continent's future representation. At the same time, through these images, he creates a tactical procedure that serves the later colonization and Christianization of America. Under the conceptual deployment of the marvelous, Columbus' letter brings together the economic, religious and imaginary aspects of the marvel coming from the historical context of the Middle Ages and the incipient early modern period. Columbus' images of America fixed a stereotypical and exoticized conception of the region. This is what the Latin American neo-baroque in the 20th century revisited in order to reflect on Latin American identity and to provide such a meaning to the conception of wonder, which would differ from the one linked to colonization in the admiral's letter.

This chapter analyzes how the images of America as a land of "marvel", "gold" and "savages" that Columbus depicted in his letter are revised with a decolonial perspective in Lezama Lima's baroque and Alejo Carpentier's "*Real Maravilloso*" (Marvelous Real). I aim to show how "wonder" intrinsically maps out a continental, historical and epistemic journey between Europe and America, and how it suggests a relation between modernity and the baroque. The notion of marvel studied in this chapter stems from 1940s/1960s Cuban literature and functions as a critical and identity-based element that decolonizes the original, exoticizing sense of the concept. Here, "marvel" works as an aesthetic that speaks about otherness. Connecting the wonder that is discussed in Columbus's letter with the Marvelous Real involves a considerable chronological leap. However, rather than focusing on a strictly defined historical period, the purpose of this article is to understand wonder as a notion that triggers an aesthetic procedure related to modernity and coloniality.

The concept of modernity and the timing of its beginning have been widely debated.[7] In this chapter I subscribe to the decolonial perspective advocated by Walter Mignolo and Enrique Dussel, according to whom modernity begins in the 15th century with the arrival of the Spaniards to the American continent.[8] This point of view regards coloniality as the other face of modernity and argues that modernity cannot be understood without the colonial system and its inherent economic, racial and systematic violence. In addition, decolonial studies have demonstrated that the historical colonial system between the 17th and 19th centuries continued after the colonies gained their independence. Far from ending with the founding of national states, colonialism has been "reproduced epistemically and pragmatically in all dimensions of life based on the naturalization of relations of domination sustained by capital".[9] This is known as the "coloniality of power", a term coined by the Peruvian sociologist Aníbal Quijano to characterize the global domination pattern of modernity/coloniality. This pattern is racially biased and justifies existing power relations.[10] Over the years, "coloniality" has been studied as a phenomenon that refers to an institutional, epistemic and gendered domination that has been internalized in Latin American societies and remains to this day.[11]

Wonder is associated with modernity/coloniality, both historically and conceptually. Firstly, it could be said that Columbus instituted wonder in the context of the New World not only through his amazement in front of natural phenomena and the quantity of available gold, but also because, for him and his contemporary Europeans, America was a utopia and the lost paradise that suddenly materialized.[12] Secondly, the marvel functions as a colonizing agent because it established the stereotypical images that were and are used to justify coloniality. Added to this, there is an economic aspect attributed to the marvel, as it materializes in gold. This sustained colonization, as the Spanish empire saw the opportunity to recover from the austere economic situation in which it found itself at the time of Columbus' enterprise. At the same time, the gold and silver extracted from the Americas constituted the beginning of early capitalism. Finally, the marvel is related to modernity/coloniality through its function as an image-making artifact.

As I mentioned above, wonder proposes a way of seeing. Thus, "if America was a crucible of modernity, it is because it was also a fabulous laboratory of images".[13] Gruzinsky refers to this as "the war of images" specifically with reference to the episode that led to the overthrow of the indigenous gods by Cortés and the subsequent fall of Tenochtitlan in 1519. The overthrow of the indigenous gods was followed by their replacement by Christian images.[14] Cortés himself referred to this as: "removing the idols and placing the images".[15] This process implied a fetishization of the indigenous gods, whose representations became a kind of collectible marvelous object.[16] By replacing them with Christian images, Europeans inaugurated iconoclasm in the new world, which, together with writing, would constitute one of the main instruments of European culture.[17] In this sense, I argue that the marvel embedded in the admiral's letter established a procedure that elaborated images and constituted imaginaries through which Columbus saw America as if through a "template".[18]

By this, I mean that Columbus already had a predetermined mold: "His template is already sketched, it has a certain shape on a translucent paper that lets us see what is behind it, but distorts it".[19] In his rhetoric, the marvel is an image that distorts.

The marvel in Columbus's letter contains a colonizing aesthetic. It is the aim of this chapter to review how the notion of the marvel underwent a form of decolonization in the 20th century with Lezama Lima's neo-baroque and Carpentier's Marvelous Real. The question of the marvelous resurfaced again in the European avant-garde at the beginning of the century but was re-signified during the 1940s and 1960s in the literary and theoretical proposal from both Cuban authors who assessed Latin American identity by re-observing the marvelous.

Columbus's letter and its images

The letter to Luis de Santángel is a brief missive sent by Columbus on 15 February 1493 to the Catholic Monarchs's Treasurer. In the letter, Columbus writes that he had reached what he believed to be the Indies and then provides a general but tactical report of what he had found in the New World.[20] According to Robert Folger, Columbus's letter responds in a self-fashioned way to the social organization system of the time: an "Economía de mercedes" ("economy of rewards"), which means claiming a reward for the services to the Crown.[21] After initial difficulties in convincing the monarchs to finance his enterprise, Columbus managed to change their minds by arguing for an expansion of the Spanish empire and finding an alternative route that would lead them to Asia without going through the Ottoman blockade. The letter served as an opportunity for Columbus to prove that it was worth the investment and that he would be able to repay or even exceed the initial expenditure. This is why his letter functioned not only as an inventory but also as a lavish portrayal of what he had found.

Besides exemplifying Columbus's rhetorical strategies, the letter to Santángel is a historical document of high value; it is the testimony of the first contact with the Americas, and it legitimizes an unlikely enterprise. This chapter addresses the interaction of all these layers and shows how in his letter Columbus established three images that constituted the rationale behind the colonization of America and remain as stereotypes today. These are: "The gold", "the good savage/cannibal" and "the marvel". Prior to the analysis of these images, it is necessary to understand the context in which they were depicted and the reasons why Columbus described America with these three concepts. For this kind of close reading of 1492, it is helpful to follow Sylvia Wynter, who, through the interrelated concepts of "subjective understanding" and "stereotyped image" analyzes the cultural and historical aspects of Columbus' arrival and the subsequent colonization of the new world.[22]

Firstly, the "subjective understanding" is a concept borrowed from the artificial intelligence theorist Jaime Carbonnell to explain how human beings narrate the world to themselves. It is a system of representation that is constituted by a series of beliefs shared within a historical community and adopted as one's own. Therefore, anything outside this subjective understanding causes fear. However, as an individual, Columbus was a figure who transcended the understanding typical to his time as he questioned the belief system (popular and Christian) that nourished this *propter nos* by imagining that there was a world beyond Cape Bojador (also known as the Pillars of Hercules, where the end of the world was supposed to be). Thus, the arrival of Columbus in America ("the Indies") destabilized the image and cartography of the world and shook this system of shared beliefs.

Subjective understanding is also influenced by the second concept: the "stereotyped image". The functions of these images were "to induce the specific mode of perception needed by a culture-specific order, and to thereby orient the prescribed behaviors needed by that order".[23] Consequently, these images determine the conception of the belief system that encompasses subjective understanding and serve as delimiters for anything that attempts to transgress. Stereotyped images order and maintain boundaries. In light of Wynter's concepts, it is possible to see a contradiction in Columbus: while he challenged the subjective understanding of his time with the stereotyped images he produced in his letter, he also controlled the effects/impact of the arrival in America. Thus, although Columbus brought a new world to light, the images he conceived of the natives were the stereotypes that maintained the order of this understanding.[24] As a result, a symbolic system of "race" was developed from here, a system which served as a basis for colonization and which, as I mentioned above, persists to this day with the "coloniality of power".

Returning to the letter, the three images that stand out in Columbus' writing (the "gold", the "good savage/cannibal" and the "marvel") constituted the stereotyped images with which the admiral kept the European-Christian subjective understanding in order. With them, he staged a new world, but one which he immediately accommodated to his belief system. Firstly, the search for gold became a delirium because, in addition to justifying the financing of Columbus's project, the Spanish empire was trying to recover economically after the expulsion of the Jews in 1492. The gold of America marked the beginning of early capitalism and the start of the plundering that, supposedly, would finance the reconquest of Jerusalem. In the letter, the admiral makes a strategic mention of the gold he had learned about, but which he had not yet found. This is why, when describing the nature around him, Columbus writes that "the number and size and wholesomeness of the rivers, most of them bearing gold".[25] The gold is represented as part of the abundance of this land and as something that would legitimize his enterprise. For this reason, at the end of the letter, he reassures that the amount of gold is certain and that he has witnesses. Furthermore, he promises that he will give the kings "all the gold they required, if they will give me but a very little of assistance".[26] Gold is decisive here, because it signifies two different things: a natural resource for America and the symbolic power granted to it by imperial forces in Spain. In this difference, we also measure the extraction, accumulation and enslavement that inaugurated early capitalism.

The second image presents an ambiguity between the good savage and the cannibal, reflecting both a fascination and a rejection of the unknown Other. The inhabitants of the New World soon became the human resource that justified the process of evangelization and the Christian mission that increased the number of subjects of the Spanish empire. For this reason, Columbus immediately made them appear as beings with potential for religious conversion, not only because of their "good heart" and "generosity" but also because they believed that he and his men were from heaven.[27] Of these good savages, he says: "I have not found in those islands any monsters, as many imagined; but on the contrary, the whole race is very well-formed".[28] This docility has to do with a kind of feminization-infantilization of the indigenous that began with Columbus and continued to serve a later colonial representation. This association suggests that everything indigenous (land, gold, bodies) can be possessed.[29] In the letter, Columbus describes the inhabitants of the New World as helpless, naive and unarmed, and expresses amazement about the fact that men had long hair like women.

The image of the passive Indian, however, is contradicted by a small, seemingly insignificant reference at the end of the letter. Columbus relates that, according to what he had heard from other groups, there seemed to be a tribe of "Caribs" who ate people.[30] This brief mention is crucial because it established a colonial imagery that later unleashed both intense theological debates about the existence of the soul of the indigenous people and an iconographic representation that reconstructed anthropophagic scenes.[31] These developments have fostered a deep rupture between "the one" and "the Other". The savage is the other face of the civilized man and, as such, represents a figure that conceals the crimes of the white Europeans.[32] In the 20th century, the Brazilian avant-garde responded to the image of the anthropophagist through Oswald de Andrade's "cultural anthropophagism".[33] With this concept, the author seeks to reverse the image of the cannibal by proposing a cultural devouring. However, both the image of the good savage and that of the cannibal reinforce a colonial perspective and an imaginary that reintegrates otherness framed in a sort of bestiary, in the best style of the books of wonders.

The third image—"marvel"—contains all the other images. The multiple significance of the marvel gives it a character marked by tactical-rhetorical, religious and economic considerations. The word "marvel" appears six times throughout the letter. Sometimes, Columbus uses it to express his surprise in front of natural phenomena: "La Spañola is a marvel; the mountains and hills and plains and fields".[34] Later in the letter, he remarks that "there are palm-trees of six species, wondrous to see for their beautiful variety". Here, wonder coincides with a way of seeing, as I discussed above with reference to the etymology of the word. The term is also used as an adverb of quantity. Here, for instance, it appears to say that the people of the island are "most wondrously timorous".[35]

In Columbus's letter the marvelous thus functions as a word to express admiration and abundance. It allows the appearance of the indescribable by giving it a frame, an empty signifier that can be occupied by images of a different nature. This is what I refer to as the distorting character. "Marvel" is, therefore, the possibility of saying everything and nothing. It is a general word that escapes the object it tries to name. Columbus made use of the marvel because he understood the wide tradition behind the concept and the social impact that could be achieved with its strategic employment. Marvel is connected to the miraculous, associated with the relics of saints and related to alchemy. Embedded in the marvel is the search for paradise, and, in the case of Columbus, this was realized with everything that he found in America. America itself was the place of wonder, it was the object of desire that materialized through gold, nature and its inhabitants.

According to Greenblatt, Columbus's wonder is a "linguistic and legal gap". In my view, it goes beyond that, as it triggers a cultural gap. Paradoxically, the wonder serves as an empty signifier and establishes the amputation of America. The marvel marks an exteriority in which everything is differentiated, foreign and other. In the past, the marvel initiated a colonial and exoticizing gaze on America and a method for "othering". In the process by which Columbus turned the New World into a marvel, we can see the ambiguity of coloniality as the other face of modernity. This goes in line with the other part of Greenblatt's definition. He states that, in Columbus, wonder works "as the agent of conversion: a fluid mediator between outside and inside, spiritual and carnal, the realm of objects and the subjective impressions made by those objects, the recalcitrant otherness of a new world and the emotional effect aroused by that otherness".[36]

The marvel, as an abstraction that establishes a boundary between what is foreign and what is our own—the marvel object, the marvel affection, the marvel Other—allows us to trace a journey. This journey spans from the rhetoric and colonial images instituted by Columbus to the decolonial aesthetics found in the literature of Lezama Lima and Carpentier, which return to the marvel. The letter to Santángel marks a precedent for subsequent encounters with the marvelous, and it had significant historical consequences for the conception of the American identity in art and literature. The marvel was redefined and endowed with meaning through the *Marvelous-Real* (Real Maravilloso) and the American neo-baroque.

The Marvelous Real and the (neo)baroque

Between 1940s and 1960s, Latin American literature expressed an ontological concern for defining the "American being".[37] It is precisely in this concern where Alejo Carpentier's Marvelous Real (*Real Maravilloso*) and Lezama Lima's neo-baroque are located.[38] Carpentier's 1948 essay "*De lo real maravilloso americano*" ("On the Marvelous Real in America") forms part of the prologue to his novel *El reino de este mundo* (The kingdom of this world). In his essay, Carpentier begins a journey from East to West, recreating Columbus's voyage in the opposite direction. On this journey, Carpentier pauses to observe and becomes amazed by the unknown, while recognizing his lack of understanding. He admits that he only feels qualified to speak of the reality he knows. But it is also through that reality that the distinction between what is "own" and what is "other" makes sense. For him, the Marvelous Real interpreted the Latin American in dialogue with the universal. One of the most common objections to both Cuban authors is that they still had a European approach to literature. Therefore, it is necessary to point out that, without denying the colonial and European past, both authors dedicated themselves to theorizing Latin American literature and identity. They understood its identity interweaving within a contradictory context in which it was always a fundamental question. The Marvelous Real does not mean a negation of the European. Decolonizing the image of the Wonder of Columbus does not mean disregarding a colonial historical past but turning the marvelous into a visual methodology that appreciates stereotypical images and reverses them.

His years in Paris kept Carpentier in contact with the artistic expressions of the moment. It was there that his literature of the Marvelous Real emerged and began to differentiate itself from the European avant-garde. Carpentier criticized surrealism for the way it prefabricated the marvelous. In his view, the surrealist notion of the marvelous was artificial. To Breton's claim that "everything marvelous is beautiful", Carpentier responded that "it is neither beautiful nor ugly; it is most of all astonishing because of its unusualness. Everything that is unusual… everything that is out of the established norms is marvelous".[39] Carpentier criticized surrealism for pursuing the marvelous *outside* reality. The European marvelous is a notion to escape from "*Unbehagene in der Kultur*" ("discomfort of culture") of which Freud would speak five years after the publication of the surrealist manifesto in 1929. Breaking free from reality and creating artistic media to achieve this was an important theme for the European avant-garde and especially for surrealism, which had been interested in the findings of psychoanalysis on the unconscious and its relation to the dream world. For Carpentier, in contrast, the marvelous "is found at every step".[40] The marvelous arises *inside* of reality, in everyday life. The marvelous is a confrontation, a learning to live with contradiction, a way of understanding reality and not escaping from it.

Besides surrealism, Carpentier also criticized Franz Roh's "*Magischer Realismus*" (1925) and his expressionist paintings, although it was the German language and the expressionist movement where the concept of the Marvelous Real originated from.[41] For Carpentier, the problem with the European interpretation was that it combined real forms in a way that did not correspond to everyday reality. Roh's compositions were improbable, the image was stopped. For Carpentier, this did not mean that the marvelous would not have existed in Europe. His main criticism was that it was "manufactured by tricks of prestidigitation, by juxtaposing objects unlikely ever to be found together".[42] For Carpentier, the marvelous could not be produced. Instead, he writes that

> (t)he Marvelous begins to be unmistakably marvelous when it arises from an unexpected alteration of reality (the miracle), from privileged revelation of reality… or an amplification of the scale and categories of reality, perceived with particular intensity by virtue of an exaltation of the spirit that leads it to a kind of extreme state ("*estado límite*"). To begin with, the phenomenon of the marvelous presupposes faith.[43]

The marvelous is, then, an unusual event. It is produced from an ambiguity of reality, and it is an exaltation, a form that leads to an extreme state. Nevertheless, even though the marvelous is part of everyday life, it is not evident. Everything Carpentier has said so far seems to be linked to a traditional view of the marvelous; the difference lies in the way he uses wonder as an aesthetic methodology, which ends up becoming an ontological question.[44]

Carpentier's view of the marvelous proposes a praxis of the gaze that culminates in an artistic praxis. Accordingly, the initial statement of his essay reflects why he conceives the "artist as detector of realities".[45] Thus, contrary to the exoticization typical of Columbus's marvel, Carpentier's marvelous is a method. It is a recognition of the multiple conditions of miscegenation, of colonial heritage in which contradictions arise and coexist. The Marvelous Real is expressed through the (neo)baroque. The (neo)baroque as a rupture emerged again in the 20th century with the literature of Lezama Lima, Carpentier and Severo Sarduy.

According to Carpentier, this "baroque" was reborn in America, but not as a historical style. During the 17th century, the historical baroque had a sumptuous splendor to it, following the European arrival in America, as the gold that was extracted from the New World played a conspicuous role in constructing the baroque ornaments. While American Baroque architecture flourished in Peru and Ecuador, in literature, one of its great exponents was the Mexican Sor Juana Inés de la Cruz. However, the American (neo)baroque should be considered a "spirit", a "human constant".[46] Thus, it appeared again during the 20th century as an aesthetic of rupture. While Carpentier believed that America had always been baroque, Lezama Lima saw in the baroque a project of modernity. For Latin America, the baroque is a form of knowledge that counterweights instrumental reason. For Lezama, for instance, baroque style has its realization through a counterpoint of images always in tension, which is related to the image that Gilles Deleuze proposes for the baroque: "the fold".[47] It is possible to illustrate this by reflecting on the folds of Bernini's Saint Theresa's dress during the 17th century. This quintessential baroque image showcases a superposition of layers in tension, concealing something beneath. American baroque with Lezama Lima and Carpentier, however, digs into these layers to show the oppositions. Unlike the European baroque, which hides, the neo-baroque unfolds and opens levels of meaning. This is not to say that Latin American baroque is not complex or that the words that it uses are organized with a clarity that does not require interpretation. In fact, the works of Lezama and Carpentier are known for their intricate narratives. What the American baroque does, as Irlemar Chiampi states, is that it "allows us to reinterpret the American experience as a dissonant modernity"[48] and with this "to make evident the contradictions of the modern experience".[49]

How is the marvelous expressed through the baroque? In his 1978 novel *El arpa y la sombra* (The Harp and the Shadow), Carpentier recreates the ambiguity of the marvelous in his personification of Columbus. On the one hand, there is a Columbus who wants to be canonized and who needs to be praised for his discovery. On the other hand, we have a Columbus who, *ad portas* of death, finds himself defeated and reflects on his actions with shame, thinking of the glory that finding a new world would have brought, but which, in the end, ends up being his punishment. Having the "wonder" in his hands is also his misery.

After returning from his first trip, Columbus narrates how he is received with honors by the monarchs, who invite him to the court. On this day, he wants to show his findings, so he prepares a great exhibition, a "tableau of wonders". The tableau functions as a theater piece, in which he both produces and performs. For the representation, he brings some indigenous people from America, but he cannot show them naked for respect to the kings. He considers dressing them like Spaniards, but regrets it, because they could look very similar to "certain sunny-skinned Andalusians".[50] Thus, with the help of a Jewish tailor, he decides to dress them with

> "(r)ed breeches sewed with little gold threads ('That's it... Perfect!' I repeated, 'as good as gold'), filled with exquisite feathers-although not from the birds on their islands-that drape gracefully down the backs of their heads" [51]

In addition to the indigenous people, Columbus also uses gold in different figures and sizes for his exhibition, but they are so small that the amount of gold seems insignificant. Of course, he could not miss a sample of the nature of the New World, some plants and animals of sizes that didn't exist in Europe.

He also had the Indians carry on their shoulders parrots that had survived the journey. However, he was a little worried that they were going to talk too much, because

> "I had given them many crumbs soaked in red wine, which made them raise such a hubbub that I was afraid they would suddenly start talking, repeating the bad words they had certainly heard on board the ship".[52]

In these examples, we see how Carpentier, in his re-elaboration of Columbus's voyage, returns to the images that Columbus had fixed on America. This time, however, everything Columbus does is not seen as a great achievement, but rather perceived as something ridiculous. Things that seemed marvelous to Columbus in the New World do not look the same when moved to another place. This is why the concept of the marvelous is empty and has to be filled with meaning. When wonder is dislocated, it seems to occupy the place of parody and therefore has to be produced in some way. This is the reason why Columbus has to make a performance and why the New World out there is completely artificial and false inside the court. Columbus had to invent the Indian, his clothes and his feathers. The good savages who had been generous are now reduced to some kind of animality. Suddenly, they do not fit on the other side of the ocean; they cannot be presented to the kings as they are. They have to be civilized, but moderately and not too much, so that they still embody the amazement of otherness. Finally, when they are presented to the kings, the natives begin to cry and beg to be returned to their place of origin.

Gold, as the promise of abundance, no longer seems to be available in large amounts. Columbus cannot give proof of the gold he had been told about. Hence, upon his return, he cannot be taken seriously. Even though he speaks of an exorbitant amount of this mineral, the court possesses more gold than what was brought by Columbus. Additionally, the exuberant nature that had astonished him because of its abundance is now diminished and withered after the long voyage. The parrots begin to vomit, and Columbus must remove them from the room. Nothing that the admiral presents is impressive, and the whole ceremony is nothing more than a performance that falls into absurdity.

Here, Carpentier shows us a damaged, broken marvel. All of its brightness is overshadowed in the narrative of the novel. When the admiral sees his theater piece getting out of hand, he removes everything he had brought from the New World. Thus, left with no other choice, he begins to talk compulsively about the travel and his heroic endeavors in order to fill with words the wonder that cannot be seen. Lacking the wonderful, he has to create it. This demonstrates how the traditional baroque discourse tries to name with more and more words something that cannot be named and, in so doing, moves further and further away from the object it is describing.

Carpentier reveals the emptiness of Columbus' wonder, which has to be filled with fantasy and amazement. As the work of the Mexican historian Edmundo O'Gorman shows, Carpentier tells us about the discursive invention of America.[53] America is invented and imagined, its wonder is placed from the outside as something foreign and empty. This is why wonder cannot be easily detected. In order to perceive wonder, it is necessary to know how to look. This is precisely what the neo-baroque does: it teaches how to look, it exercises a gaze that unfolds each fold. While the neo-baroque is aware of the fact that words can overshadow, confuse and turn into a meaningless noise, it works on this tension and plays with revelation and concealment. It works with likelihood, with something that is and is not at the same time. In the neo-baroque, we live with the difficulty of naming the marvel and accepting the contradiction involved. This is also present in Bolívar Echeverría's philosophy, where he speaks of an American baroque ethos that makes "the unlivable livable". The ethos of an époque should be understood as "the prevailing response of that time to the need to overcome the unbearably contradictory character of its specific historical situation".[54] The Baroque emerged as a response to the codigophagy of the 17th century, when the European code consumed the American code through miscegenation. Thus, the only subsistence option left to the American was to transform itself, to "invent a life within death".[55] Echeverría affirms that the modern way of being is baroque because it "allows us to live the destruction of the qualitative, by allowing access to the creation of another dimension".[56] A baroque way of life consists of a creative attitude that constructs ways of inhabiting where there are only possibilities of death. This narrative "possibility" was precisely what Lezama and Carpentier sought to achieve in their work.

However, the difficulty of naming, this tension between representation and reality makes sinister (uncanny) the expression of the Marvelous Real in the baroque. Irlemar Chiampi explains this when she states that Carpentier's writing is aphasic:

> "In Carpentier, the impact of the nameless produces, notoriously, the effect of the uncanny. His aphasic rhetoric is indesignable… because of the proliferating baroque, which never reaches the full naming- it turns the American real into a sinister entity".[57]

Somehow, the Marvelous Real evidences the *Un-heimlich* (uncanny)—as Chiampi asserts with reference to Freud—because it is found in the daily, in the familiar (*heimlich*), but at the same time it arises from the unusual. The baroque is "*unheimlich*" in its reactionary condition to reason. In the 17th century, it was irrational when reason was subversive. In the 20th century, it was irrational when it opposed itself to institutionalized reason and portrayed capitalist and instrumental modernity.[58] The baroque is "*unheimlich*" because it speaks of a loss and the impossibility of recovering it. The neo-baroque and its expression in the Marvelous Real oscillate between the limits of the rational and the irrational. It challenges an aesthetic that allows narrating a contradictory identity. This difficulty of naming that becomes sinister is taken up again by the baroque in its connection with modernity.

The uncanny condition is evident in the prologue to *The kingdom of this world* (1949), in which Carpentier presents his theory of the Marvelous Real. Referring to the Haitian revolution, Carpentier narrates the metamorphosis of Mackandal, a black slave condemned to the stake who, by transforming himself into a mosquito, is able to escape his condemnation: "and Mackandal, transformed into a buzzing mosquito, would light on the very tricorne of the commander of the troops to laugh at the dismay of the whites".[59]

What is *unheimlich* here is that Mackandal's transformation and salvation is seen only by the slaves. The whites, on the contrary, see him die and are terrified by the indolence of the others who watch the event impassively. The baroque implies the visualization of other realities. It implies seeing beyond and traces a possibility in the abyss. It is not in vain that in his essay, Carpentier argues that the Marvelous Real presupposes a faith. The sinister reality that the baroque and the Marvelous Real bring us closer to is part of the re-signification of the wonder. The marvelous does not imply an easy solution or the vision of a sweetened reality but proposes to look at all its implications.

Conclusion

To conclude, I would like to come back to the initial question and observe how, after travelling to the marvelous-real and the American baroque, we return to the image of Columbus' letter as a place of wonder in order to look at it from another perspective and change the focus through which the Other is exoticized. In addition to it being a form of expression, the Marvelous Real questions an exoticizing episteme and creates an aesthetic of otherness. Thus, the marvel in itself, in the work of Carpentier and Lezama, traces a decolonizing journey in which it goes from being an empty signifier to one in which it endows reality with familiarity and strangeness. The wonder's travel speaks of its narrative consequences and its epistemic transformations. Wonder is a concept appropriated by Latin American literature to speak of contradictory realities. The marvel of both Latin American authors puts into perspective historical, social and economic contexts that intertwine and present marvel as an aesthetic methodology, which is put into practice in Latin American literature. The marvel is a notion that unleashes aesthetic procedures related to modernity and coloniality.

Notes

1 Sievers 2007, 771.
2 Daston and Park 1998, 16.
3 Daston and Park 1998.

4 Daston and Park 1998.
5 See *The travels of Marco Polo* (13th century), which had a great influence on Columbus' conviction that he had reached the Indies. Also, Mandeville's *The book of marvels and travels* was fundamental to spread the imaginary of the marvelous in the Middle Ages.
6 Daston and Park 1998, 67.
7 Echeverría 2000, 17, dates the beginning of modernity to the 11th century and to the development and premeditated use of instruments copied from nature. According to Koselleck 1989, 30, modernity starts from the 17th century due to the change of attitude toward the perception of time and how the future appears as an expectation.
8 Mignolo 2001, 39.
9 Gaviria and Nina 2023, 180. The translation from German to English is mine.
10 Quijano 2014, 780.
11 See, for example, *The coloniality of knowledge: Eurocentrism and social sciences: Latin American perspectives*, edited by Edgardo Lander,CLACSO, 2000, which discusses various aspects of the coloniality of power. For the gender perspective, see Lugones 2008, who argues that besides issues relating to race it is necessary to also include the question of gender in our discussions on coloniality. This idea has been essential for the entire intellectual production of decolonial feminism.
12 Todorov 1987 points out that in his vision of America as the lost paradise, Columbus places himself in the figure of Adam by renaming the islands with names that allude to kings and religion and by changing the spelling of his name to resemble that of Christ (1982, p. 34).
13 Gruzinsky 2019, 12.
14 Gruzinsky 2019., 48.
15 Cortés, *Cartas y relaciones*, 106.
16 Around 1496, before Cortés, Columbus and the friar Juan Ramón Pané, whom Columbus had commissioned to investigate the antiquities of the Indians (see his *Relación acerca de las antigüedades de los indios*), found a group of objects that the Indians worshipped and called by a generic name: Zemiés. These objects were of various kinds and some were more venerated than others. These objects "were endowed with political functions, therapeutic and climatic properties, the Zemies have sex, speak and move" (Gruzinsky 2019, 20). Pané called such objects idols, a word which in his account was sometimes replaced by "demon" and which Columbus himself was careful to use, because he refused to attribute idolatry to the Indians.
17 Gruzinsky, 2019, 12.
18 Gaviria and Nina 2023, 189. The translation from German is mine.
19 Gaviria and Nina 2023. The translation from German is mine.
20 Columbus based his knowledge on the Ptolemaic model, which miscalculated the circumference of the Earth to be smaller than what it really was. This is why Columbus believed he had reached the Indies. However, there is a scientific background to his voyage, supported by the Renaissance and humanist mentality. By the time Columbus set out on his voyage, technological advances such as the compass, the printing press and gunpowder already existed, bolstering Columbus's belief in a voyage of such dimensions.
21 Folger 2011, 22.
22 Wynter 1995, 15, 34.
23 Wynter 1995., 20.
24 Wynter 1995., 39.
25 Colombus, *Letter to Luis de Santangel*, 5.
26 Colombus, *Letter to Luis de Santangel*, 15.

27 Colombus, *Letter to Luis de Santangel*, 6.
28 Colombus, *Letter to Luis de Santangel*. 13. In the last part, Spanish version states: "*es toda gente de muy lindo acatamiento*", of which a closer translation would be "They are all very well-behaved people". This is directly related to obedience and submissiveness.
29 Segato 2016, 19, states that the female body appears as the first colony, as the first occupied territory. Without this original event of colonization, the subsequent conquest (of Spain over America) could not have been thought of. In this sense, if we think of Jan van der Straet's engraving "Allegory of America" (ca.1587–89), we can see how the new continent was represented with the image of a naked and defenseless woman. See the engraving in the MET museum's web page: https://www.metmuseum.org/art/collection/search/343845.
30 Segato 2016, 14.
31 The most famous theological debate was the one between Bartolomé de las Casas and Ginés de Sepúlveda in Valladolid (1550). On the debate, see Sylvia Wynter, "New Seville and the Conversion Experience of Bartolomé de Las Casas: Part One and two", *Jamaica Journal* 17:2, May and August (1984): 25–32; 46–55. For the anthropophagic acts of the indigenous people, see the engravings of Theodor de Bry (although his images are based on the chronicles of Hans Staden (1557) and Jean de Léry (1578), De Bry never travelled to America).
32 Bartra 2011, 15, argues "that savage men are a European invention that essentially obeys the internal nature of Western culture". Bartra dismantles the discourse in which the savage is assimilated only to the indigenous groups in America.
33 Oswald de Andrade published his "Manifesto Antropofagista" in 1928 in the first issue of the *Revista de Antropofagia*, directed by Antonio de Alcântara and Raul Bopp.
34 Colombus, *Letter to Luis de Santangel*, 4–5.
35 Colombus, *Letter to Luis de Santangel*, 6.
36 Greenblatt 1991, 75.
37 Chiampi 2011, 50, discusses the context of a generation of Latin American essayists between 1920 and 1960, who influenced the writing and thinking of the two Cuban authors. The question that marked this generation was directed toward the cultural identity of the continent, the search for which became an ontological concern to define itself intellectually.
38 References to the marvelous real often trigger discussions about "magical realism". Although they are two completely different subjects, the concept of "magical realism" has been used since the 1960s and 1970s as a marketing strategy around the boom publishing phenomenon. This meant that the concept lost the purpose that motivated it in its origins and was imbued with an aura of exoticism that does not correspond to its aesthetic purposes.
39 Carpentier 1995b, 101.
40 Carpentier 1995a, 88.
41 Guerrero 2000, 54–56. The concept of "magical realism" was introduced in 1948 by Arturo Uslar Pietri in the Hispanic American literary field to talk about the work of the Guatemalan Miguel Ángel Asturias and to analyze aspects of the Venezuelan short story.
42 Carpentier 1995a, p. 85.
43 Carpentier 1995a, 86.
44 The ontological question in Latin American literature appears already in the 1920s, as Chaimpi points out in his essay "La historia tejida por la imagen" (History woven by the image). There are also many reflections on this in Carpentieren's texts, as discussed in this chapter. For Lezama, see: *La expresión Americana* (1957); *Las eras imaginadas* (1971).
45 Carpentier 1995a, 76.

46 Carpentier 1995b, 94.
47 Lezama Lima 2011, 218.
48 Chiampi 2000, 18.
49 Chiampi 2000.
50 Carpentier 1992, 99.
51 Carpentier 1992, 100.
52 Carpentier 1992, 101.
53 See *The Invention of America* by O'Gorman (1958). More than the historical fact itself, the author explores the discovery of America in an abstract sense and analyzes how the images, narratives and beliefs about the new continent were part of a narrative construction, an invention in which Europe projected itself.
54 Echeverría 2000, 89.
55 Echeverría 2006, 214.
56 Echeverría 2006.
57 Chiampi 2000, 115. Translation is mine.
58 Chiampi 2000.
59 Carpentier 1957, 51.

Bibliography

Primary sources

Columbus, Christopher. Letter to Luis de Santangel. In *Select Letters of Christopher Columbus*, edited and translated by R.H. Major, 1–18. London: Hakluyt Society, 1870.

Cortés, Hernán. *Cartas y relaciones de Hernán Cortés al emperador Carlos V*, Paris Maisonneuve 1866.

Mandeville, John. *Book of Marvels and Travels*. Oxford: Oxford University Press, 2012.

Literature

Bartra, Roger. *El mito del salvaje*. México, D.F: Fondo de Cultura Económica, 2011.

Breton, André. *Manifiestos del surrealismo*. Edited by Aldo Pelegrini. Buenos Aires: Argonauta, 1992.

Carpentier, Alejo. *The Kingdom of this world*. Translated by Harriet de Onís. New York: The Noonday Press, 1957.

Carpentier, Alejo, *The Harp and the Shadow*. Translated by Carol and Thomas Christensen. London: André Deutsch Limited, 1992.

Carpentier, Alejo. "On the Marvelous Real in America". In *Magical Realism: Theory, History, Community*, edited by Lois Parkinson Zamora and Wendy B. Faris, 75–88. Durham: Duke University Press, 1995a.

Carpentier, Alejo. "The baroque and the Marvelous Real in America". In *Magical Realism. Theory, History, Community*, edited by Lois Parkinson Zamora Ans Wendy B. Faris, 89–108. Durham: Duke University Press, 1995b.

Chiampi, Irlemar. "La historia tejida por la imagen". In *La expresión americana*, Ensayos completos III: Lezama Lima, José, 49–78. Almería. Confluencias, 2011.

Chiampi, Irlemar. *Barroco y modernidad*. México: Fondo de cultura Económica, 2000.

Daston, Lorraine; Park, Katharine. *Wonders and the Order of Nature 1150-1750*. New York: Zone Books, 1998.

Gaviria Isabel and Nina, Fernando. "Bild, Körper und affektives Verstehen. Der Traumals dekoloniale Praxis in 'El sueño del Pongo'". In *Leteinamerikanische Literaturatheorie*, edited by Fernando Nina, 179–208. Berlin: Tranvía, 2023.

Greenblatt, Stephen. *Marvelous Possessions: The Wonder of the New World.* Oxford; Chicago: The University of Chicago Press, 1991.

Gruzinsky, Serge. *La guerra de las imágenes. De Cristóbal Colón a "Blade Runner" (1492–2019)*. Translated by Juan José Utrilla. México: Fonde de Cultura Económica, 2019.

Guerrero, Gustavo. "Uslar Pietri, cronista del realismo mestizo", *Cuadernos Hispanoamericanos*, 605 (2000): 53–62.

Echeverría, Bolívar. *La modernidad de lo barroco*. México, D.F: Ediciones Era, 2000.

Echeverría, Bolívar. *Vuelta de siglo*. México, D. F: Ediciones Era, 2006.

Echeverría, Bolívar. *¿Qué es la Modernidad?* México: Universidad Nacional Autónoma de México, 2009.

Folger, Robert. *Writing as Poaching: Interpellation and Self-Fashioning in Colonial relaciones de méritos y servicios, The Medieval and Early Modern Iberian World.* Leiden; Boston: Brill, 2011.

Koselleck, Reinhart. *Vergangene Zukunft: zur Semantik geschichtlicher Zeiten*. Frankfurt am Main: Suhrkamp, 1989.

Lezama, Lima. "La curiosidad barroca". In *La expresión americana*, edited by Chiampi, Irlemar, 214–241. Almería: Confluencias, 2011.

Lugones, María. "Colonialidad y género", *Tabula Rasa* 9 (2008): 73–101.

Mignolo, Walter. "La colonialidad: la cara oculta de la modernidad". In: *Cosmópolis: el trasfondo de la Modernidad*, 39–49. Barcelona: Península, 2001.

Quijano, Aníbal. "Colonialidad del poder, eurocentrismo y América Latina". In *Cuestiones y horizontes: de la dependencia histórico-estructural a la colonialidad/descolonialidad del poder*, 777–832. Buenos Aires: CLACSO, 2014.

O Gorman Edmundo. *La invención de América*. México D.F.: Fondo de Cultura Económica, (1995) [1958].

Segato, Rita. *La guerra contra las mujeres*. Madrid: Traficantes de sueños, 2016.

Sievers, Julie. "Literatures of Wonder in Early Modern England and America," *Literature Compass* 4, no. 3 (2007): 766–783.

Todorov, Tzvetan. *La conquista de América: el problema del otro*. México: Siglo Veintiuno editores, 1987.

Wynter, Sylvia. "1492 a New World View". In *Race, Discourse, and the origin of the Americas*. Edited by Vera Lawrence; Rex Nettleford, 5–57. Washington; London: Smithsonian Institution Press, 1995.

INDEX

Pages followed by "n" refer to notes.

FSC
www.fsc.org
MIX
Papier | Fördert gute Waldnutzung
FSC® C083411

Kolibri 360 GmbH
Ferdinand-Jühlke-Straße 7
99095 Erfurt, Deutschland
produktsicherheit@kolibri360.de